Culture & Leisure Services
Red Doles Lane
Huddersfield, West Yorks. HD2 1YF

More praise for
The Making of the Fittest

"So many things in this book that I haven't read before. . . . I recommend it to everybody who wants to understand more because it is written in such simple yet detailed language, quite a pleasure to read."
—Ira Flatow, NPR "Science Friday"

"Carroll is an adept and wide-ranging writer. . . . Reading *The Making of the Fittest* is like spending a few hours with an extremely knowledgeable and enthusiastic dinner companion."
—Steve Olson, *Washington Post*

"Conveying the excitement of current research while also providing a firm foundation of why we know what we know is a rare gift. In *The Making of the Fittest*, Carroll offers a graceful and insightful view of the explanatory power of evolution."
—Douglas Erwin, *American Scientist*

"Excellent. . . . Carroll's book will certainly help the public to understand evolution more clearly."   —Brian Charlesworth, *Nature*

"With fervor and clarity, Carroll amasses a glut of facts to refute the twisted logic of the anti-Darwinist camp."—Josie Glausiusz, *Discover*

"Carroll is not as rude and impatient as I am with those Luddites of the life sciences who, at the first hint of disagreement with their beliefs, stick their fingers in their ears and sing, 'La la la.' His book is friendly and charming and, by its conclusion, quietly devastating in its condemnation of stupidity. Read it and shout, 'Hallelujah!' "
—Peter Birnie, *Vancouver Sun*

"Students and teachers of biology will particularly benefit from his readable treatment of the evolutionary process. An essential addition to every school, public, and academic library." —*Library Journal*

"A fast-paced look at how DNA demonstrates the evolutionary process. . . . Carroll offers some provocative and convincing evidence." —*Publishers Weekly*

"Of all the scientists in the world today, there is no one with whom Charles Darwin would rather spend an evening than Sean Carroll."
—Michael Ruse, author of *The Evolution-Creation Struggle*

"Sean Carroll is our plain-spoken emissary from the next great revolution in biology. His earlier book, *Endless Forms Most Beautiful*, provides a fine introduction to the amazing field called Evo Devo. Now in *The Making of the Fittest* he offers something even more fundamental—glimpses of what molecular genetics is revealing about the process and course of evolution. This book is fascinating, lucid, surprising, and (in the truest sense) essential."
—David Quammen, author of *The Reluctant Mr. Darwin* and *The Song of the Dodo*

"Crime-scene investigators love DNA evidence. It can close a case that couldn't be solved any other way. This fascinating book presents the DNA evidence for evolution, and anyone who reads it should agree that it's an open-and-shut case. I hope *The Making of the Fittest* will make a difference in the one and only place where Darwin's case is still seriously debated: the court of public opinion."
—Jonathan Weiner, Pulitzer Prize–winning author of *The Beak of the Finch*

"Sean Carroll's gift as a writer is the way in which he invites his readers to see science from the inside and the ease with which he explains the scientific wonders of cutting-edge research in biology. Always captivating, always accessible, *The Making of the Fittest* is a book for all readers, and one that fulfills Darwin's promise that the science of evolution would ultimately illuminate every aspect of the study of life itself. In Carroll's hands, it surely does."

—Kenneth R. Miller, author of *Finding Darwin's God*

ALSO BY SEAN B. CARROLL

Endless Forms Most Beautiful:
The New Science of Evo Devo

The Making of the Fittest

· · · · · · · · · · · · · · · · ·

DNA AND THE ULTIMATE FORENSIC
RECORD OF EVOLUTION

Sean B. Carroll

WITH ILLUSTRATIONS BY
Jamie W. Carroll and Leanne M. Olds

Quercus

First published in Great Britain in 2008 by

Quercus
21 Bloomsbury Square
London
WC1A 2NS

A CIP catalogue reference for this book is available
from the British Library

ISBN: 978 1 84724 4 765

10 9 8 7 6 5 4 3 2 1

Printed and bound in Great Britain by Clays Ltd, St Ives plc.

For Joan H. Carroll and the late J. Robert Carroll.

Thanks for the DNA—and all of my mutations.

Contents

The Making of the Fittest

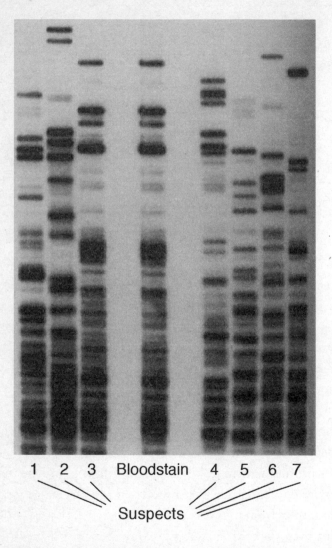

1   2   3   Bloodstain   4   5   6   7

Suspects

Forensic DNA analysis. The banding pattern in each lane is a "profile" of the DNA of crime scene samples and potential suspects. The bloodstain DNA matches that of suspect 3 and no others. *Photograph copyright Cellmark Diagnostics.*

Beyond Any Reasonable Doubt

. . . .

Facts do not cease to exist because they are ignored.
—Aldous Huxley

IN 1979, DIANNA GREEN, NINE MONTHS PREGNANT, was severely beaten and her unborn child died from the trauma. Even though she had amnesia and could not spell her name at trial, she testified that her husband, Kevin Green, had inflicted her injuries. Green was convicted of murder and attempted murder.

In 1996, workers at the California Department of Justice laboratory processed the seventeen-year-old murder scene evidence for DNA analysis. By comparing the samples' DNA profile with those of Green, and in a newly created offender database, they discovered that the original sample and samples from four other murders matched that of a different man, Gerald Parker, who was then in prison on a parole violation. Confronted with the DNA evidence, Parker confessed to the crimes (and was subsequently sentenced to death), and Green was freed from prison after serving sixteen years for a crime he did not commit.

More accurate and rigorous than fiber or fingerprint analysis, and far more reliable than eyewitness testimony, DNA analysis can provide conclusive proof about who was or was not at the scene of a crime. The authority of DNA evidence, and many cases similar to Green's, have led to a revolution in the criminal justice system and a vast increase in the use of DNA testing to both convict the guilty and exonerate the innocent. Many crimes that would have been unsolvable in the past are now solved routinely, including "cold cases" several decades old. The number of exonerations is also growing. The Innocence Project, an organization that provides pro bono representation for DNA-based appeals, reports more than 150 exonerations over the past thirteen years, including many individuals freed from death row.

The power of DNA testing extends far beyond criminal justice. The determination of paternity is now definitive, and testing for carriers of genetic diseases is now routine, thanks to DNA science. But there is one arena where that power is not yet widely appreciated: in what one might call the philosophical realm.

Just as the sequence of each individual's DNA is unique, the sequence of each species' DNA is unique. Every evolutionary change between species, from physical form to digestive metabolism, is due to—and recorded in—changes in DNA. So, too, is the "paternity" of species. DNA contains, therefore, the ultimate forensic record of evolution.

This presents an interesting irony. Juries and judges are relying on DNA evidence to determine the liberty or incarceration, and life or death, of thousands of individuals. And it appears that citizens in the United States are 100 percent supportive of this development. Yet, in the court of public opinion, some 50 percent or more of the U.S. population still doubts or outright denies the reality of biological evolution. We are clearly more comfortable with DNA's applications than with its implications.

Over a century ago, William Bateson began one of the first important books on evolution to appear after Darwin's time with the exhortation: "If facts of the old kind will not help, let us seek facts of a new kind. That the time has come for some new departure most naturalists are now I believe beginning to recognize."

WITH DNA science penetrating so many facets of everyday life, it is again time for a new departure and to seek facts of a new kind. My goal in this book is to present a body of new facts about evolution gathered from DNA evidence. Over the past few years, biology has gained unprecedented access to a vast amount of DNA evidence from all kinds of organisms, including humans and our closest relatives. In just twenty years, the amount of DNA sequences in our databases has grown 40,000-fold, with the vast majority of that coming in this new century. To put that number in perspective, in 1982 our total knowledge of DNA sequences from all living species amounted to fewer than one million characters. If printed onto pages like those you are reading, that amount of text would fit easily into one volume about the size of this book. If all of the DNA text that we now have was printed into volumes and stacked, they would reach more than double the height of the 110-story Sears Tower in Chicago. This library of life is growing by more than 30 stories per year.

Inside these books is the raw DNA code for the making of all sorts of bacteria, fungi, plants, and animals. The decoding of these texts, composed of almost infinite permutations of sequences of just four letters—A, C, G, and T—now presents one of the greatest opportunities in the history of evolutionary biology. Biologists are mining this rich new resource to investigate and solve some of the most fascinating mysteries in natural history and to reveal, in unprecedented detail, how all sorts of important traits have evolved in nature. In this book I will tell the story of how the new science of *genomics*—the comprehensive and, most especially, the *comparative* study of species DNA— is profoundly expanding our knowledge of the evolution of life.

Genomics allows us to peer deeply into the evolutionary process. For more than a century after Darwin, natural selection was observable only at the level of whole organisms such as finches or moths, as differences in their survival or reproduction. Now, we can *see* how the fittest are *made*. DNA contains an entirely new and different kind of information than what Darwin could have imagined or hoped for, but which decisively confirms his picture of evolution. We can now identify specific changes in DNA that have enabled species to adapt to changing environments and to evolve new lifestyles.

This new level of understanding provides more than just definitive forensic evidence, it includes some surprises that expand our picture of evolution. For example, in the DNA record of every species, we find fossil genes. These are bits of DNA text that were once intact and used by ancestors, but that have fallen into disuse and decay. These relics are an entirely new source of insights into traits and capabilities that have been abandoned as species evolved new lifestyles.

The DNA record also reveals that evolution can and does repeat itself. Similar or identical adaptations have occurred by the same means in species as different as butterflies and humans. This is powerful evidence that, confronted with the same challenges or opportunities, the same solution can arise at entirely different times and places in life's history. This repetition overthrows the notion that if we rewound and replayed the history of life, all of the outcomes would be different.

DNA evidence is also revolutionizing the study and understanding of human origins and early civilization. While the sequencing of the human genome has grabbed most of the headlines, it is the decoding of the genes and genomes of other primates and mammals that enables us to interpret the meaning of the human text. Our genes contain telltale clues to both how we are different and how we evolved to be so. Many genes bear the scars of natural selection—of the battles our ancestors fought with the germs that have plagued human civilization for millennia.

I wrote this book with a variety of readers in mind. First, for those with a keen interest in natural history, I will roam the planet to show how many fascinating species have adapted to boiling-hot springs, caves, jungles, lava formations, the deep ocean, and other remarkable places. There is grandeur in this new knowledge of how changing one or a few letters in simple code can dramatically change the form or physiology of complex organisms. Second, for students and teachers, I have focused on what I believe are the best examples that illustrate the key elements of the evolutionary process while reinforcing and expanding our awe for the amazing diversity and adaptability of life. Most of the stories I tell are not yet in textbooks, but many will become key chapters in evolutionary science. And third, for those trying to sift through the rhetoric and pseudoscience of evolution's opponents, I have provided some background to understand the tactics and arguments used to doubt and deny evolutionary science, and plenty of scientific evidence to vaporize those arguments.

The new DNA evidence has a very important role beyond illuminating the process of evolution. It could be decisive in the ongoing struggle over the teaching of evolution in schools and the acceptance of evolution in society at large. It is beyond ironic to ask juries to rely on human genetic variation and DNA evidence in determining the life and liberty of suspects, but to neglect or to undermine the teaching of the basic principles upon which such evidence, and all of biology, is founded. The anti-evolution movement has relied on entirely false ideas about genetics, as well as about the evolutionary process. The body of new evidence I will describe in this book clinches the case for biological evolution as the basis for life's diversity, beyond any reasonable doubt.

Bouvet Island. Photograph by Ditlef Rustad, Norvegia expedition, 1928.

Bouvet Island, as seen and photographed by Ditlef Rustad on the 1928 *Norvegia* expedition. *Photograph from* Scientific Results of the Norwegian Antarctic Expeditions, 1927–28, *published by I. Kommisjon Hos Jacob Dybwad of Oslo, 1935.*

Chapter 1

Introduction:
The Bloodless Fish
of Bouvet Island

· · · · · · · · · · · · ·

When we no longer look at an organic being as a savage
looks at a ship, as at something wholly beyond his com-
prehension; when we regard every production of nature
as one which has had a history; when we contemplate
every complex structure and instinct as the summing up
of many contrivances, each useful to the possessor, nearly
in the same way as when we look at any great mechanical
invention as the summing up of the labor, the experience,
the reason, and even the blunders of numerous workmen;
when we thus view each organic being, how far more
interesting, I speak from experience, will the study of
natural history become!

—Charles Darwin,
On the Origin of Species (1859)

IT MAY BE THE MOST REMOTE PLACE ON EARTH.
Tiny Bouvet Island is a lone speck in the vast South
Atlantic, some 1600 miles southwest of the Cape of Good
Hope (Africa) and almost 3000 miles east of Cape Horn

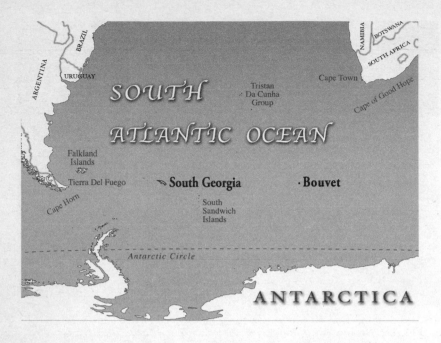

Fig. 1.1. **Map of the Southern Ocean.** *Drawn by Leanne Olds.*

(South America) (figure 1.1). The great Captain James Cook, commanding the HMS *Resolution*, tried to find it on his voyages through the Southern Ocean in the 1770s, but failed both times. Covered by an ice sheet several hundred feet thick that ends in sheer cliffs, which in turn drop to black volcanic beaches, and with an average temperature below freezing, it still doesn't get many visitors.

Fortunately, for both my story and natural history, the Norwegian research ship *Norvegia* made it to Bouvet Island in 1928, with the principal purpose of establishing a shelter and a cache of provisions for shipwrecked sailors. While on Bouvet, the ship's biologist, Ditlef Rustad, a zoology student, caught some very curious-looking fish. They looked like other fish in most respects—they had big eyes, large pectoral and tail fins, and a long protruding jaw full of teeth. But they were utterly pale, almost transparent (figure 1.2; color plates A and B).

FIG. 1.2. **An icefish.** *Photograph permission of Italian Antarctic Program, PNRA.*

When examined more closely, Rustad noticed that what he called "white crocodile fish" had blood that was completely colorless.

Johan Ruud, a fellow student, traveled to the Antarctic two years later on the factory whaling ship *Vikingen*. He thought the crew was pulling his leg when one flenser (a man who stripped the blubber and skin from the whale) said to him, "Do you know there are fishes here that have no blood?"

Playing along, he replied, "Oh, yes? Please bring some back with you."

A good student of animal physiology, Ruud was perfectly sure that no such fish could exist, as textbooks stated firmly that all vertebrates (fish, amphibians, reptiles, birds, and mammals) possess red cells in their blood that contained the respiratory pigment hemoglobin. This is as fundamental as, well, breathing oxygen. So when the flenser and his friends returned from the day's efforts without any *blodlaus-fisk* (bloodless fish), Ruud dismissed the idea as shipboard lore.

Ruud returned to Norway the following year and mentioned the tale to Rustad. Much to his surprise Rustad told Ruud, "I have seen such a fish," and showed him the photographs he had taken on his expedition.

Ruud heard nothing more about the bloodless fish for twenty years. Then, another Norwegian biologist returned from an Antarctic expedition with white-blooded fish from a different location. His curiosity reawakened, Ruud began to ask other colleagues voyaging to the Antarctic to be on the lookout for what the whalers called "devilfish" or, because of their near transparency, "icefish." Finally, Ruud returned to the Antarctic himself in 1953, almost twenty-five years after his first journey, with the hope of catching and studying these fish and resolving the mystery of their blood.

He set up a makeshift laboratory on South Georgia Island (the island to which explorer Ernest Shackleton rowed in 1916 in order to save the stranded crew of the *Endurance*). He promptly received a few precious specimens and carefully analyzed their odd blood. His findings, reported in 1954, are still a shock for any biologist reading them for the first time. The fish completely lacked red blood cells, the pigmented oxygen-carrying cells that, until the discovery of these Antarctic icefish, had been found in every living vertebrate. Indeed, no other case of bloodless vertebrate has ever been discovered outside of the fifteen or so species of icefish now known.

Red blood cells contain large amounts of the hemoglobin molecule, which binds oxygen as blood cells circulate through the lungs or gills, and then releases it as red cells circulate through the rest of the body. The hemoglobin molecule is made up of a protein called globin and a small molecule called heme. The red color of blood is due to the heme that is buried in the hemoglobin molecule and actually binds the oxygen. We would, and do, die without red cells (anemias are conditions of low red cell numbers). Even close relatives of the icefish, such as Antarctic rock cod and New Zealand black cod, are red-blooded.

The existence of these remarkable fish provokes many questions. Where, when, and how did they evolve? What happened to their hemoglobin? How can the fish survive without it or red blood cells?

The typical place one would begin to explore the origin of a species would be the fossil record. However, that is completely lacking for these fish and their relatives. And, even if we had fossils, we would not be able to tell, from the remnants of their bones, what color their blood was and when it changed. But, there is a record of the history of icefish that we can access—in their DNA.

The clear, stunning answer to the question of what happened to their hemoglobin came from the study of icefish DNA more than forty years after Ruud first sampled their blood. In these amazing fish, the two genes that normally contain the DNA code for the globin part of the hemoglobin molecule have gone extinct. One gene is a molecular fossil, a mere remnant of a globin gene—it still resides in the DNA of the icefish, but it is utterly useless and eroding away, just as a fossil withers upon exposure to the elements. The second globin gene, which usually lies adjacent to the first in the DNA of red-blooded fish, has eroded away completely. This is absolute proof that the icefish have abandoned, forever, the genes for the making of a molecule that nurtured the lives of their ancestors for over 500 million years.

What would provoke such a dramatic rejection of a way of life that serves every other vertebrate on the planet?

Necessity and *opportunity*, both of which sprang from dramatic, long-term changes in ocean temperature and currents.

Over the past 55 million years, the temperature of the Southern Ocean has dropped, from about 68 degrees F to less than 30 degrees in some locales. About 33 to 34 million years ago, in the continual movement of the Earth's tectonic plates, Antarctica was severed from the southern tip of South America, and became completely surrounded by ocean. Ensuing changes in ocean currents isolated the waters around the Antarctic. This limited the migration of fish populations such that they either adapted to the change, or went extinct (the fate of most). While others vanished, one group of fish exploited the changing ecosystem. The icefish are a small family of species, within the larger suborder Notothenioidae, that altogether contains about 200 species and now dominates the Antarctic fishery.

The low temperature of Antarctic waters presents some great challenges to body physiology. Like the oil in my car during a Wisconsin winter, the viscosity of body fluids increases in the subfreezing Antarctic water temperatures, which would make them difficult to pump. Antarctic fish, in general, cope with this problem by reducing the number of red cells per volume of circulating blood. Red-blooded Antarctic fish have hematocrits (the percentage of their blood volume made up of red cells) of around 15 to 18 percent, while we have hematocrits of about 45 percent. But the icefish have taken this to the extreme, by eliminating red blood cells altogether, and allowing their hemoglobin genes to mutate into obsolescence. These fish, whose blood is so dilute that it contains just 1 percent cells by volume (all white cells), literally have ice water in their veins! How does this creature cope with the absence of life-sustaining hemoglobin?

It is clear now that the loss of hemoglobin has accompanied a whole suite of changes in the fish that allows it to thrive at below-freezing temperatures. One of the important differences between warm and cold water is that oxygen solubility is much greater in cold water. The frigid ocean is an exceptionally oxygen-rich habitat. Icefish have relatively large gills and have evolved a scaleless skin that has unusually large capillaries. These two features increase the adsorption of oxygen from the environment. Icefish also have larger hearts and blood volumes than those of their red-blooded relatives.

Icefish hearts differ in another obvious and profound way—they are often pale. The rose color of vertebrate hearts (and skeletal muscles) is due to the presence of another heme-containing, oxygen-binding molecule, called myoglobin. Myoglobin binds oxygen more tightly than hemoglobin and sequesters it in muscles so that it is available upon exertion. The muscles of whales, seals, and dolphins are so rich in myoglobin that they are brown in color; their high myoglobin allows these diving mammals to stay submerged for long periods. But myoglobin is not a stand-in for the absence of hemoglobin in icefish. It is absent from the muscles of all icefish and the hearts of five species (hence their paleness). The myoglobin protein is encoded by a single gene in vertebrates. Analysis of the DNA of pale-hearted icefish revealed that their myoglobin gene is mutated—an insertion of five additional letters of DNA has disrupted the code for making the normal myoglobin protein. In these species, the myoglobin gene is also on its way to becoming a fossil gene. The fishes' many cardiovascular adaptations are providing sufficient oxygen delivery to body tissues in the complete absence of two fundamental oxygen-carrying molecules.

Life in very cold waters demands yet further accommodations, and the unmistakable evidence of evolutionary change is found in many more places in icefish DNA. Even basic structures in each cell must be modified to adapt to life in the cold. For example, microtubules form a critical scaffold or "skeleton" within cells. These structures are involved in cell division and movement as well as in the formation of cell shapes. With so many important jobs to do, the proteins that form the microtubules are among the most faithfully preserved not just in all vertebrates, but in all eukaryotes (the group including, among others, animals, plants, and fungi). In mammals, microtubules are unstable at temperatures below 50 degrees F. If this were the case in Antarctic fish, they would certainly be dead. Quite to the contrary, microtubules of Antarctic fish assemble and are very stable at temperatures below freezing. This remarkable change in microtubule properties is due to a series of changes in the genes that encode components of the microtubules, changes that are unique to cold-adapted fish, both icefish and their red-blooded Antarctic cousins.

There are many more genes that have been modified so that all sorts of vital processes can occur in the subfreezing climate. But adaptation to cold is not limited to the modification of some genes and the loss of others; it has also required some *invention*. Foremost among these is the invention of "antifreeze" proteins. The plasma of Antarctic fish is chock-full of these peculiar proteins, which help the fish survive in icy waters by lowering the temperature threshold at which ice crystals can grow. Without them, the fish would freeze solid. These proteins have a very unusual and simple structure. They are made up of 4 to 55 repeats of just three amino acids, whereas most proteins contain all 20 different types of amino acids. Since warm-water fish have nothing of this sort, the antifreeze genes were somehow invented by Antarctic fish. Where in the world did antifreeze come from?

Chi-Hing Cheng, Arthur DeVries, and colleagues at the University of Illinois discovered that the antifreeze genes arose from part of another, entirely unrelated gene. The original gene encoded a digestive enzyme. A little piece of its code broke off and relocated to a new place in the fish genome. From this simple nine-letter piece of DNA code, a new stretch of code evolved for making the antifreeze protein. The origin of the antifreeze proteins stands out as a prime example of how evolution works more often by tinkering with materials that are available—in this case a little piece of another gene's code—rather than by designing new things completely from scratch.

As a resident of a cold climate, I have to admire the icefishes' grit and ingenuity. We take various measures to keep our cars running on subzero Wisconsin days, but the icefish has managed to change its whole engine *while the car was running*. It invented a new antifreeze, changed its oil (blood) to a new grade with a remarkably low viscosity, enlarged its fuel pump (heart), and threw out a few parts along the way—parts that had been used in every model of fish for the past 500 million years.

The DNA record of icefish, and of all other species, is a whole new level of evidence of the evolutionary process. It allows us to see beyond the visible bones and blood, directly into the fundamental text of evolution. The making of the extraordinary icefish illustrates the ordinary, if somewhat messy, course of the making of the fittest at the DNA level. Icefish evolved from warm-water, red-blooded ancestors ill suited to life in the cold. Their adaptation to the changing environment of the Southern Ocean was not a matter of instant design, nor just a one-way "progressive" process. It was an improvised series of many steps, including the invention of some new code, the destruction of some very old code, and the modification of much more.

By comparing the states of genes in different icefish, their closest red-blooded relatives, and other Antarctic fish, we can see that certain changes occurred at different stages in icefish evolution. All 200 or so Antarctic notothenioid species have antifreeze genes, so that was an early invention. So, too, were the modifications of microtubule genes. But only the fifteen or so icefish species have fossil hemoglobin genes. This means that the hemoglobin genes must have been abandoned by the time the first icefish evolved. Furthermore, while some icefish can't and don't make myoglobin, others do. This reveals that the changes in the myoglobin genes are more recent than the origin of icefish, and that the use (or disuse) of myoglobin is still evolving. By examining other DNA sequences from each species, it is possible to map these events onto the timeline of the geology of the South Atlantic—with the origin of the Antarctic Notothenioidae occurring about 25 million years ago, and the origin of the icefish only about 8 million years ago (figure 1.3). The DNA record tells us that the icefish crossed the divide between a warm-water, hemoglobin-dependent lifestyle, and a very-cold-water, hemoglobin- (and for some myoglobin-) independent lifestyle in many steps, not in a leap.

The DNA record of the many modifications accumulated by the icefish in the long course of its descent from red-blooded, warm-water ancestors vividly demonstrates the two key principles of evolution—natural selection, and descent with modification—first articulated by another zoology student, Charles Darwin, who journeyed around the South Atlantic a century before Rustad and Ruud. In order to fully appreciate the power of this new DNA record I am going to describe throughout this book, and its place in the larger picture of the evolutionary process, it is important to refamiliarize ourselves with these two principles and their initial statement in *On the Origin of Species*.

Darwin Redux

Darwin boarded the HMS *Beagle* in December 1831, at age twenty-two, for what would eventually become a five-year voyage that circumnavigated the globe. The bulk of the voyage was spent in and around

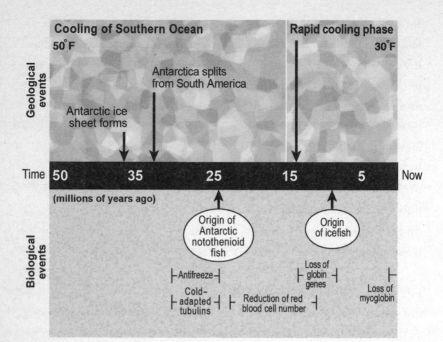

FIG. 1.3. **A timeline of icefish evolution.** (Top) The changing geology of Southern landmasses brought about major changes in ocean currents and temperatures over the past 50 million years. (Bottom) One large group of fish, called the notothenioid fish, adapted to lower temperatures by evolving antifreeze, cold-stable tubulins, and a lower hematocrit. Eventually, globin genes were fossilized in the common ancestor of bloodless icefish. *Drawing by Leanne Olds.*

South America, as the obsessive Captain Robert Fitzroy charted and recharted rivers and harbors. For Darwin, the animals, plants, fossils, and geology of this vast continent set in motion what would emerge more than twenty years later as *On the Origin of Species*, the opening lines of which read:

> When on board H.M.S. *Beagle* as naturalist, I was much struck with certain facts in the distribution of the organic beings inhabiting South America, and in the geological relations of the present to the past inhabitants of the continent. These facts . . . seemed to

throw some light on the origin of species—that mystery of mysteries, as it has been called by one of our greatest philosophers. On
my return home, it occurred to me, in 1837, that something might
perhaps be made out on this question by patiently accumulating
and reflecting on all sorts of facts which could possibly have any
bearing on it. After five years of work I allowed myself to specu-
late on the subject, and drew up some short notes; these I enlarged
in 1844 into a sketch of the conclusions, which then seemed to me
probable; from that period on to the present day I have steadily
pursued the same object. I hope that I may be excused for entering
on these personal details, as I give them to show that I have not
been hasty in coming to a decision.

His "abstract" ran 502 pages and sold out in one day, on November
24, 1859.

"How extremely stupid not to have thought of that!" exclaimed the
great biologist Thomas Huxley after reading *On the Origin of
Species.*

Contrary to most popular notions, it was not the idea of evolution
that was novel in Darwin's book. That possibility had been floating
around for many decades, indeed, in Darwin's own family. His grand-
father, Erasmus Darwin, put forth a theory of evolution in *Zoonomia,
or the Laws of Organic Life* (1794).

Nor was it the mere idea of species changing that prompted
Huxley's reaction. Rather, it was the power, yet intuitive simplicity, of
two ideas—"Descent with Modification" and "Natural Selection"—
as the description of and mechanism for life's evolution.

Darwin drew an analogy between the selection for variation in the
domestication of animals and the struggle for existence among the far
more numerous offspring produced in the wild than are able to thrive:

Can it, then, be thought improbable . . . that other variations useful in some way to each being in the great and complex battle of
life, should sometimes occur in the course of thousands of genera-

tions? If such do occur, can we doubt (remembering that many more individuals are born than can possibly survive) that individuals having any advantage, however slight, over others, would have the best chance of surviving and of procreating their kind? On the other hand, we may feel sure that any variation in the least degree injurious would be rigidly destroyed. *This preservation of favorable variations and the rejection of injurious variations I call Natural Selection* [emphasis added].

—Ch. IV, *On the Origin of Species*

Darwin then leaped to the bold conclusion that this process would connect all life's forms via their descent from common ancestors:

Several classes of facts . . . seem to me to proclaim so plainly, that the innumerable species, genera, and families of organic beings, with which this world is peopled, have all descended, each within its own class or group, from common parents, and have all been modified in the course of descent.

—Ch. XIII, *On the Origin of Species*

And then, even bolder:

Therefore I should infer from analogy that probably all the organic beings which have ever lived on this earth have descended from one primordial form, into which life was first breathed.

—Ch. XIII, *On the Origin of Species*

This is the essence of Darwinian evolution—that natural selection for incremental variation forged the great diversity of life from its beginning as a simple ancestor. Simple logic, scientific immortality. No wonder Huxley was chiding himself.

But there was much more in *On the Origin of Species* than these few conclusions (some of which Alfred Russel Wallace had independently reached as a result of his studies in South America and the

Malay Archipelago). Darwin brought *evidence*. Mounds and mounds of observations, fact upon fact, ingenious experiments, clever analogies, and twenty years of finely crafted argument.

The esteem we biologists have for Darwin is manifold. Sure, *On the Origin of Species* is the most important single work in biology. Darwin's "long argument" is brilliantly constructed, supported by a dazzling breadth of facts, and the product of a heroic individual effort. It is also very readable today, with its passion still resonant. But the full body of his contributions filled many books, from insights into the building of coral reefs, to the importance of sexual selection, and the biology of orchids, barnacles, and much more. It just dwarfs what the merely talented or industrious might achieve.

So why have his great ideas endured such a struggle?

Seeing the Steps

Darwin understood all too well, and therefore correctly anticipated, most of the objections that could or would be raised against his ideas. Many of the attacks, of course, were from those who found Darwin's view of life's history repulsive and demeaning on nonscientific grounds. Most scientists fairly readily accepted the reality of evolution, that is, that species did change. But even Darwin's supporters had difficulties with the *how*—with the mechanism he proposed.

Questioning the *how* was quite understandable. I believe that most people, scientists or laypeople, initially struggle to get their heads around Darwin's picture of natural selection, or what also became known as "the survival of the fittest." (An interesting note: Darwin did not coin that famous phrase, the philosopher Herbert Spencer did. It did not appear in *On the Origin of Species* until the fifth edition of 1869, at Wallace's suggestion.) Darwin's process of evolution involved three key components—variation, selection, and time. Each of these presented some conceptual or evidential problems, and all were potential sources of incredulity. Darwin was asking his readers, in essence, to imagine how slight variations (whose basis was unknown and invisible) would be selected for (which occurred by a process that was also invisible and not measurable) and accumulate over a period of time that was beyond human experience. Darwin understood the difficulty:

> the chief cause of our natural unwillingness to admit that one species has given birth to other and distinct species, is that we are always slow in admitting any great change *of which we do not see the steps* [emphasis added]. The difficulty is the same as that felt by so many geologists when Lyell first insisted that long lines of inland cliffs had been formed, and great valleys excavated, by the slow action of the coast waves. The mind cannot possibly grasp the full meaning of the term of a hundred million years; it cannot add up and perceive the full effects of many slight variations, accumulated during an almost infinite number of generations.
>
> —Ch XIV, *On the Origin of Species*

The eminent biologist and writer Richard Dawkins points out that the concept of natural selection *is* simple but deceptively so: "It is almost as if the human brain were specifically designed to misunderstand Darwinism, and to find it hard to believe." The individual components of chance (in producing variation) and selection (in determining which variants succeed) can be so easily misunderstood or confused. The role of chance is often inflated (sometimes deliberately so by opponents of evolution) to mean that evolution occurs completely at random, and that order and complexity arise all at once and at random. This is not at all the case. Selection, which is not random, determines what chance occurrences are retained. It is the *cumulative* selection ("adding up," in Darwin's terms) of variations that forges complexity and diversity, over periods of time that we humans just don't grasp very well. Natural selection even strained Darwin's supporters. They just couldn't see how selection could be powerful enough to "see" and to accumulate slight variations.

It wasn't until some fifty years after *On the Origin of Species* that biologists finally appreciated the interplay of chance, selection, and time in concrete terms. It turns out that a little bit of everyday math, the kind we use to calculate probabilities in a casino or in a lottery, and to calculate interest on savings and loans, finally convinced biologists (including some prominent doubters) that natural selection was, at least in theory, strong enough and fast enough to account for evolution.

But math goes only so far. Just as for Johan Ruud and the tale of the bloodless fish of the Antarctic, for most of us believing and understanding is a matter of seeing. We want to see the stuff responsible for evolution. We want to be able to see, measure, and retrace the steps taken in evolution between one species and another.

Now, after 140 years, we can do just that.

The DNA Record of Evolution

Each step in evolution, we now know, is taken and recorded in DNA. Every change or new trait—from the antifreeze in the bloodstream of Antarctic fish, to the beautiful colors of an alpine wildflower, to our large brain-packed skulls—is due to one or more (sometimes many, many more) stepwise changes in DNA that are now traceable. Some steps are tiny, just a single change in one letter of one gene's code. Others are much larger, involving the birth (and death) of entire genes or blocks of genes in one leap.

We can track these changes because of the explosive increase in our knowledge of species genes and genomes (the entire DNA content of a species). From just a trickle of the small genomes of bacteria and yeast several years ago, the large genomes of complex creatures such as the chimpanzee, dog, whale, and various plants are being revealed at a torrential pace. The unique DNA sequence of each species is a complete record of the present. It is an inventory of all the genes used to build and operate that creature.

The DNA record is also a window into the recent and the deep past. When the first genome of a member of some group is determined, that pioneer paves the way for much faster analysis of its relatives. By comparing genes and genomes between relatives of different ranks, we can pinpoint important changes and spot the mark of natural selection. The view can be as humbling as it is exciting. We can peer back a few million years to track the changes that took place in the evolution of the line that led to us from our common ancestor with the chimpanzee, our closest relative on the planet. We can look back 100 million years or so to see what gave rise to the differences between marsupial and placental mammals. We can even glimpse before the dawn of animals and find hundreds of genes in simple, single-celled organisms that evolved more than two billion years ago and still carry out the same jobs in our bodies today.

The ability to see into the machinery of evolution transforms how we look at the process. For more than a century, we were largely restricted to looking only at the outside of evolution. We observed external change in the fossil record and assessed differences in anatomy. But before this new molecular age there was no way to make genetic comparisons between species. We could study the reproduction and survival of organisms and infer the forces at work. However, we had no concrete knowledge of the mechanism of variation or the identity of the meaningful differences between species. Yes, we understood that the outcome was the survival of the fittest, but we did not know *how the fittest are made*. Just as for any work of human creation, we so much better understand how complex things have come to be—cars, computers, spacecraft—when we understand how they are made, and how each new model is different from its predecessors. We are no longer savages staring at passing ships.

The focus of this book will be to peer into the DNA record to see how evolution works. Along the way we will explore how some of the most interesting and important capabilities of some fascinating creatures arose. The book is organized into three main parts. I would like to think of them as being like the three parts of a good and memorable meal—a little bit of preparation, plenty of food, and some meaningful conversation. First, in order to prepare for the meal, I want to take some care in explaining the main ingredients of evolution—variation, selection, and time—so that we fully appreciate how they interact in the making of the fittest.

The late Nobel laureate Sir Peter Medawar once remarked that "the reasons that have led professionals without exception to accept the hypothesis of evolution are in the main too subtle to be grasped by laymen."

I don't believe that this is true. However, if it is at all true, this is a failure on the part of scientists to clearly explain the power of natural selection, compounded by time, to make all things great and small—from whales to bloodless icefish.

To redress this shortcoming, I will explain the everyday math of evolution (chapter 2). This is the best way to get a good feel for the power of natural selection and to vanquish some of the misleading arguments against the probability of events in evolution. This simple math is generally not explained in popular accounts of evolution. It is important, however, for grasping not just the plausibility of natural selection, but also the real-world interplay of chance, time, and selection. I know, you are saying, "Math?! Forget it." Don't worry. At the very minimum, that chapter might help you become a better gambler or investor.

The main body of the book will be a six-course meal, served in six chapters. The focus of each chapter will be on how the new DNA record reveals a particular aspect of evolution, with new kinds of evidence that neither Darwin nor his mathematically gifted disciples could have dreamed of.

I will begin by illustrating how the DNA record documents the processes of natural selection and descent with modification on a vast geological timescale. I will show unimpeachable evidence of how natural selection acts to remove, in Darwin's words, injurious change (chapter 3). This evidence is manifest in the form of genes that have been preserved across kingdoms of life for two *billion* years or longer. The text of these "immortal" genes is stuck "running in place" under the conservative surveillance of natural selection. Immortal genes are

more than just hardy stalwarts against the steady onslaught of mutation over eons of time, they are key pieces of evidence for the descent
of all living species from common ancestors and they provide a new
means of reconstructing early events in life's evolution.

I will then turn to the fundamental question of how species acquire
entirely new abilities and fine-tune existing talents (chapter 4). I will
focus almost exclusively on a set of the most exquisite examples of this
idea, all concerning the origin and evolution of color vision in animals.
The possession and tuning of this sense is central to animal lifestyles,
and how they find food, mates, and one another in daylight, darkness, or
in the deep blue sea. The steps to the acquisition and fine-tuning of color
vision at the DNA level are especially well understood and demonstrate
the action of natural selection on the text of evolving genes.

These examples of evolution in nature are convincing demonstrations of specific episodes and modes of evolution. In many ways they
confirm a body of theory that is many decades old. But the DNA
record would be relatively anticlimactic if it did not contain some surprises, some information that we did not anticipate but that, once
revealed, yields new insights and new ways of seeing into the evolutionary process. It has handed up some real gems.

While the focus of much of the study of life's history has centered
on the traditional fossil record, biologists have revealed in DNA a new
kind of fossil record—of fossil genes (chapter 5). Just as sedimentary
rocks contain a record of ancient forms, no longer alive, all species
DNA contains genes, sometimes numbering in the hundreds, that are
no longer used and are in various states of decay. Fossil genes, like
those I described in the icefish, are telltale clues to past capabilities,
and to shifts in species' ways of living from those of their ancestors.
Our fossil genes reveal a lot about how we are different from our
hominid ancestors.

The most profound surprise of all, though, is how evolution repeats
itself (chapter 6). When species that have independently gained or lost
similar traits are compared, we often find that evolution has repeated
itself, at the level of the same gene, sometimes right down to the very

same letter in the code of the same gene. In some cases, the same genes have become fossilized in different species. This is remarkable evidence that, in the great arc of time, different species, including those belonging to entirely different taxonomic groups, will respond in the same way to a particular selective condition. The repetition of evolution is so pervasive that we are forced to change our thinking about the uniqueness of past events. The new DNA record tells us that the probabilities are in favor not only of species coming up with particular changes in DNA, but of multiple species coming up with the same particular solutions again and again.

The repetition of evolution is not limited to the distant past or to obscure species—it is occurring in our own flesh and blood (chapter 7). Our species has been shaped by the physical environment and the pathogens we encounter. With some ancient foes, such as malaria, we are still locked in evolutionary arms races, and the scars of these battles are evident in our genes. I will explain how the process of natural selection has shaped our genetic makeup and has great implications for human biology and medicine.

The vast body of evidence I highlight in these five chapters leaves no doubt as to the ubiquity of natural selection, or its swiftness in acting on even very small differences among individuals. Yet, ever since Darwin, the most difficult aspect of the evolutionary process to grasp has been the cumulative power of natural selection in shaping the evolution of complex structures. For more than a century, detailed knowledge of the formation or history of complex organs and body parts was far out of reach.

In the final course of the main meal, I will describe recent insights into the making and evolution of complexity (chapter 8). I will emphasize how understanding the process of development reveals how complex structures are built and how comparing the development of structures of different degrees of complexity reveals how such structures evolved. The DNA record contains key insights into how complexity and diversity have evolved through the use of ancient body-building genes.

Seeing and Believing:
Why Evolution Matters

The real-time observation of the evolutionary process and the revelations of the rich and ancient DNA record set the stage for the after-dinner conversation. In the final two chapters of the book, I will confront contemporary and historical issues surrounding disbelief and acceptance of the facts of evolution and I will underscore the importance of applying evolutionary knowledge in the real world. We can learn much about the nature of the opposition to, or doubts about, evolution from previous episodes of institutional ignorance and public opposition to science—to Galileo, Pasteur, and even the science that proved DNA was the basis of heredity. The facts of astronomy, microbiology, and genetics were resisted in certain quarters until the tangible, visible evidence was overwhelming. This new DNA record cannot be argued away. The facts of evolution are overwhelming and still growing.

This book is what a critic might call "genocentric," in that it places so much emphasis on events at the level of DNA. I confess my presentation is genocentric, but my defense is that the stories I will tell are selected for their power in illustrating the adaptability of species to diverse, and, often, quite extreme habitats.

This new understanding of how the fittest are made expands our wonder at the processes that have shaped life's amazing diversity— from ancient microbes that live in boiling water to fish that breathe without hemoglobin, birds and butterflies that see colors invisible to us, and apes that write books. It also reveals why and how the "fittest" is a conditional, if not precarious, status.

The everyday math of evolution and the DNA record of life tell us that natural selection acts only on what is useful for the moment. It cannot preserve what is no longer used, and it cannot predict what will be needed in the future. Living for the moment has the dangerous disadvantage that if circumstances change more rapidly than adaptations can arise, faster than the fittest can be made, populations and species are at risk.

History shows that as circumstances have changed, globally or locally, many eras' fittest have been replaced. The fossil record is paved with creatures—trilobites, ammonites, and dinosaurs, to name just a few—of once very successful groups that evolution has left behind. The icefish have made a remarkable evolutionary journey in adapting to the changing Southern Ocean, but theirs may well be a one-way trip. Having abandoned one mode of living, they have lost capabilities that cannot be recovered. And their future is certainly in doubt.

Ditlef Rustad accidentally discovered icefish in nets he used to haul up krill, a two- to three-inch-long crustacean that is at the very center of the Antarctic food web. In late 2004, biologists studying data collected by nine countries over forty Antarctic summers reported that Antarctic krill stocks have declined by 80 percent since the 1920s. Krill feed on phytoplankton and algae that depend upon sea ice, which is shrinking, and krill are in turn eaten by squid, sea birds, whales, seals, . . . and icefish. The air temperature in the Antarctic Peninsula has risen by 4 to 5 degrees F in the last fifty years, and the water temperature of the Southern Ocean is projected to rise by several degrees over the next century. If that happens, it is very likely that most cold-adapted species will not be able to adapt to such rapid changes in temperature and food availability, and that part of the enormous and important Antarctic fishery will collapse, taking the icefish with it.

Knowledge of evolutionary biology is therefore no mere academic pursuit, nor is the acceptance of its facts a matter that should be open to political or philosophical debate.

Sir Peter Medawar also stated that "the alternative to thinking in evolutionary terms is not to think at all." That is an alternative our species can no longer afford.

DOMESTIC PIGEONS:

J. Carroll

Darwin's pigeons. The great variety of fancy pigeons derived from the rock pigeon was used by Darwin to illustrate the power of selection upon variations.

Illustrations are from The Variation of Animals and Plants Under Domestication, *Vol. 1 (London: John Murray, 1868). Montage by Jamie Carroll.*

Chapter 2

The Everyday Math of Evolution: Chance, Selection, and Time

. . . .

The whole of science is nothing more than a refinement of everyday thinking.

—Albert Einstein

EVERY FEW MONTHS OR SO, RADIO AND TV STATIONS will report that the jackpot in the multistate Powerball lottery has gone unclaimed and grown to some huge sum. This prompts a rush to purchase tickets for a chance to become filthy rich, and swells the jackpot even more. Many folks who live in nonparticipating states and don't ordinarily play the lottery will drive a fair distance and shell out quite a few bucks, perhaps figuring that it is not worth the trouble to buy a chance for $40 or $50 million, but for $200 million, now that's some real money!

California State University Professor Mike Orkin points out that if a person drives ten miles to buy a ticket, he or she is about sixteen times more likely to get killed in a car crash on the way than to win the jackpot. Wait a minute, you say; that may be for one ticket, but they're buying a lot of tickets—surely, that improves the odds. It does, but Orkin notes that a person who buys fifty tickets a week will win the jackpot on average about once every 30,000 years.

We sure do have some very funny ideas and attitudes about statistics and probabilities. And these ideas don't end with the lottery.

Stories of shark attacks always draw news coverage (not to mention spawning several movies), but the fact that the incidence of fatal shark attacks is about 1 per 300 million people per year (in the United States) doesn't seem to reduce our fear and morbid fascination. And if sharks aren't scary enough, there are mountain lions. The chance of a fatal attack happening in California, where there are growing numbers of mountain lions, is 1 in 32 million per year. Now compare that with the chances of dying of a dog bite—almost fifty times higher, at 1 in 700,000 per year—yet we surround ourselves with these lovable, slobbering killers!

Something in our nature makes us believe in our chances of beating the long odds against a great event, but we fret over even more remote odds against a tragedy, while ignoring more immediate and greater dangers. Psychology and statistics clearly do not inhabit the same parts of our brains.

I raise these examples of probabilities because evolution does involve some elements of chance. This is a major source of both doubt and confusion. Some people look at the order of nature and the marvelous ways species are adapted to their surroundings, like the icefish in the freezing waters of the Antarctic, and can't believe that this could spring from any process that involved an element of chance. Rather, they conclude that the odds are stacked against nature inventing anything new, useful, or complex. Overcoming this doubt requires understanding the interplay between chance, selection, and time. In this chapter, I will show that understanding evolution—meaning change over time—boils down to the same kind of thinking and mathematics we use (or should use in the case of the lottery!) to calculate the probabilities of everyday events.

Albert Einstein, when asked "What is the most powerful force in the universe?," replied, "Compounding interest." If Einstein had been just a little more clever, he would have said, "Natural selection"; each of these forces derives its power from the same mathematical principle. That principle might be stated simply: even though one starts with a small number (e.g., one's money in the bank) and even though the rate of increase in this number is seemingly modest per year (e.g., the interest the bank provides), given sufficient time, the growth from that initial number is dramatic after many years of compounding.

In the case of evolution, the "small number" is the number of individuals in a population with a trait and the "seemingly modest rate" is the small selective advantage that trait confers on those individuals that carry it. As I will discuss in detail shortly, the "sufficient time" for evolution is far less than one might think. While the time required for a trait to become prevalent in a population is greater than one lifetime, it is often no more than a few hundred generations. This is a blink of an eye in geologic time. This point was not appreciated until many years after Darwin formulated his theory of natural selection. The implication of this simple fact (based on simple math) is profound: it tells us that small differences among individuals, when compounded by natural selection over time, really do add up to the large differences we see among species.

Overcoming Doubts with Pigeons and Rats

First-time readers of *On the Origin of Species* may expect to be greeted with a dazzling parade of life's diversity or a sizzling narrative of human origins. They find neither. In chapter I of the most important book in all of biology, we get . . . pigeons.

That's right. After a five-year voyage around the world and more than twenty years of work and writing, Darwin opened his life's opus with English pigeons.

It was the first of many brilliant masterstrokes.

Before explaining natural selection and the descent of all species from common ancestors, he chose to explain the ideas of selection and descent in the more familiar forms of pigeon breeds.

Darwin himself was a pigeon expert. He explained at the outset, "Believing that it is always best to study some special group, I have, after deliberation, taken up domestic pigeons. I have kept every breed which I could purchase or obtain, and have been most kindly favoured with skins from several quarters of the world."

Pigeons taught Darwin about the interplay of variation and selection, and convinced him that natural selection on slight variations could, over time, account for the large differences among species.

Darwin pointed out that the varieties of pigeon were so markedly different from one another that if shown to an ornithologist and told that they were wild birds, they would each be ranked as well-defined species. But Darwin deduced correctly that they were all descended from the rock pigeon. He then applied his insights from pigeons to all of Nature.

Naturalists and breeders were misled by appearances into thinking that every kind of domestic breed (cattle, sheep, etc.) came from separate ancestors. Darwin wrote, "When I first kept pigeons and watched the several kinds, knowing well how true they bred, I felt fully as much difficulty in believing that they could ever have descended from a common parent, as any naturalist could in coming to a similar conclusion in regard to the many species of finches." His explanation for the lack of appreciation of the effect of selection was simple: "from long-continued study they [breeders] are strongly impressed with the differences between several races . . . yet they ignore all general arguments, and *refuse to sum up in their minds slight differences accumulated during many successive generations* [emphasis added]."

Darwin knew a wide community of pigeon fanciers with whom he shared knowledge on how much time was required to change traits by selective breeding. He notes that "that most skillful breeder, Sir John

Sebright, used to say, with respect to pigeons, that 'he could produce any given feather in three years, but it would take him six years to obtain head and beak.'"

Darwin was convinced of the power of natural selection over time. But even his strongest advocates had reservations.

The main sticking point was whether natural selection was efficient enough to act on small differences between individuals, or whether selection could work only on large differences. Darwin's greatest ally, biologist Thomas Huxley, believed in selection, to be sure. But Huxley found it difficult to explain the gaps between existing species and those in the fossil record as a result of natural selection acting on small continuous differences over long periods of time. Huxley preferred to think of selection acting on "saltations," which are large discontinuous differences among individuals. Huxley's favorite examples were humans and animals with extra digits. If these could arise fully formed in one generation, then the evolution of differences in digit number between species was, in Huxley's view, more readily explained by saltations than the gradual evolution of digits. Huxley held this view to his death. The question of whether natural selection was powerful enough to shape the gradual evolution of complex structures would be passed to a new generation of biologists. And for a while, things didn't look so good for Darwin.

Huxley and Darwin went to their graves completely ignorant of the mechanisms of heredity. The first rules of inheritance were discovered by the Augustinian monk Gregor Mendel in the course of breeding experiments he conducted on pea plants in the late 1850s and early 1860s (ironically, exactly at the time of publication of *On the Origin of Species*). But while Mendel was aware of Darwin, the great naturalist was never aware of Mendel's work, although the German journal in which it was described was available in Britain. It wasn't until 1900, some thirty-four years after publication and sixteen years after Mendel's death, that the scientific world took note.

One biologist who seized upon Mendel's work was William

Bateson, a Cambridge University naturalist. Bateson was seeking the laws of variation and he wrote a large book on all sorts of large, discontinuous variations found in nature. This was the foundation for his belief that selection acted on large differences among individuals, and that Darwin's picture of evolution occurring in small increments was wrong.

In Mendel's work, Bateson thought he had found the evidence to clinch his view. Mendel showed that several traits in the pea plant were inherited in a simple fashion such that the difference between pea shape or color was determined by single units (we now call these units genes). For Bateson, this was hard evidence that evolution acted on large, discrete differences such as wrinkled or smooth shape, green or yellow color—and not on anything in between. The new Mendelian evidence widened the gap between doubters and supporters of natural selection. Mendel's laws were clearly correct, so how could this impasse persist and what discoveries turned the tide back in favor of Darwin?

The turning point came, in the sort of irony that is so common in science, when the doubters considered more evidence. The story is a perfect echo of T. H. Huxley's admonition that "science warns me to be careful how I adopt a view which jumps with my preconceptions, and to require stronger evidence for such beliefs than for one to which I was previously hostile."

The discovery of Mendelian genetics energized all sorts of research programs, including experiments to improve animal breeds. One of the most important figures in this area was William Castle of Harvard University, who promptly embraced Mendelian inheritance and Bateson's view of discontinuous variation as the material for evolution. But Castle was to reverse his support of Bateson's view in relatively short order.

Castle's reversal was caused by the results of a series of breeding experiments carried out with rats over many generations. Castle and other biologists initially believed selection could not change a charac-

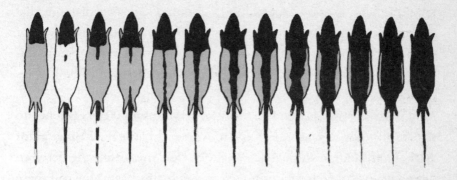

FIG. 2.1. **Selection on coat color in Castle's rats.** The extent of dark pigmentation or "hood" on these rats was modified by selective breeding beyond the limits of the patterns of starting parental lines. This was important evidence for the power of selection. *From W. E. Castle and J. C. Phillips (1914), Carnegie Institution of Washington Publication no. 195.*

ter beyond the original limits of variation of the character. Castle worked with "hooded" rats, so called because the dark fur pattern extended over the head and shoulders like a hood. He and his students found that they could produce, by repeated rounds of selective breeding, wholly new grades of fur patterns. Some patterns were intermediate between the patterns of the original breeding stocks, but others were more extreme, *beyond* the limits of the original variation present (figure 2.1). Castle realized that many genes were modifying the fur patterns and creating a continuous gradation of variation. His selective breeding scheme was acting on combinations of variants of these genes. He concluded, contrary to his original view, that selection on small degrees of continuous variation was indeed sufficient for evolution.

Castle's experiments and consequent reversal of opinion were but one line of evidence that shifted the prevailing view of evolution back toward Darwinian natural selection. In addition to experimental evidence, there also emerged a whole new approach to evolution, natural selection, and genetics—mathematics.

The Algebra of Evolution

R. C. Punnett, another early geneticist who was strongly opposed to the Darwinian view, somewhat inadvertently launched the sort of mathematical analysis that would firmly underpin natural selection. Punnett was interested in mimicry in butterflies, in which butterfly species that are palatable to birds and living in the same area evolve wing color patterns similar to that of an unpalatable species. Wanting to know how quickly selection would act to cause some trait to spread through a butterfly population or to be eliminated from it, he asked H. T. J. Norton, a mathematician, for help with the calculations.

Norton worked the numbers and found, much to the surprise of Punnett and many others, that selection and evolution were potentially much faster than expected: "Evolution, in so far as it consists of supplanting one form by another, may be a very much more rapid process than has hitherto been suspected, for natural selection, if appreciable, must be held to operate with extraordinary swiftness where it is given established variations with which to work."

The key word here is "suspected." Until Norton crunched the numbers, the time frame of selection sweeping through a population or species was foggy at best.

What Norton did was simply ask: Given some initial frequency of a trait in a population, how long would it take to increase or decrease that frequency under different *rates of selection*? The basis of Norton's calculations was straightforward, and that is where the analogy to compounding interest comes in. His question is very similar to "Given some initial amount of money, how does that amount change over time at different rates of interest?"

Those of us old enough to have retirement accounts, or fortunate enough to have some savings, should be familiar with the power of compounding. A quantity of anything—money, people, fish—grows exponentially when its increase is proportional to the quantity present. For money, the key to the rate of growth is the interest rate or rate of return. An investor earning 7 percent compounded will see her wealth double about every 10 years, while another investor earning only 1 percent compounded will see it double about every 70 years. After 70 years, that difference in interest rate means that seven doublings will occur for the first investor versus only one doubling for the other. The growth in money will be $2 \times 2 \times 2 \times 2 \times 2 \times 2 \times 2$ or 128-fold for one investor versus just 2-fold for the other, a 64-fold difference in wealth. That 6 percent difference sure adds up.

In biology, exponential growth cannot occur unabated because organisms die, and resources are finite. Darwin famously pointed out that a pair of elephants that produce six offspring in their 60-year life span would have 15 million descendants in just 500 years, even with mortality figured in. But the limited land, food, and water available means that all organisms live in competition. Competition constrains the absolute growth of populations, but it is the critical setting for natural selection. Wherever there is competition (which is everywhere) and heritable variation, selection operates.

Biologists calculate the power of selection in terms of *selection coefficients*, which are analogous to interest rates (and abbreviated as s). This coefficient indicates the incremental difference in relative reproductive success and survival between individuals with a trait and those without it. If, for example, the presence of a trait conveys a small advantage to individuals such that they produce 101 viable offspring and those lacking it produce 100 offspring, this is a 1 percent advantage (a 1 percent rate of compounding) and the value of s is positive 0.01. If there is a disadvantage of having a certain trait, say 99 offspring are produced instead of 100, s is then negative 0.01. These positive or negative values of selection coefficients are an indicator of *fitness*, which is a relative, not an absolute measure.

Just as Einstein and investors appreciate the power of compounding interest, Norton helped biologists see the power of natural selection. For example, under a modest selective advantage of just 0.01, Norton calculated that a dominant trait would increase in frequency from just 8 individuals in 1000 to greater than 90 percent of individuals in only 3000 generations. If the selective advantage was 10-fold greater ($s = 0.1$), this time was cut to just 300 generations. With many species having generation times of a year or less, these figures impressed many biologists. Much more mathematical work followed, particularly that of J. B. S. Haldane, who, along with R. A. Fisher and Sewall Wright, developed a host of formulas for understanding the relationship between evolution, selection, and time under a wide variety of different conditions.

I have been describing change in the frequency of some trait, but the compounding power of natural selection also applies to the rate of change in traits themselves. Consider the dimensions of any physical trait—the height of a plant or the length of an animal. We appreciate that there will be variation in these dimensions in any wild population. Now, suppose there is a selective advantage to the taller or longer individuals of each generation. If the rate of change per generation is just 0.2 percent, which would be 2 millimeters on a 1-meter-high plant or 1-meter-long animal, this would be imperceptible generation to generation. But, in just 200 generations, their height or length would increase by 50 percent.

These calculations demonstrate the potential power and speed of natural selection. Now, what do we know about it in reality?

Natural Selection in the Wild

Selection is much easier to see in a math formula than it is to see in the wild. In addition to the difficulty of controlling conditions, there are two major factors one can immediately appreciate that make it difficult to measure. The first is time. If changes are measurable only over periods of time that are longer than naturalists or researchers have the opportunity to record, then it would seem we are out of luck. The second challenge is the number of measurements required. The data sample must be large in order to detect subtle selective advantages or disadvantages.

This latter difficulty is a fact of probability and statistics. If the relative fitness of two forms of a species differs by a small percentage, one must count a large number of individuals over time in order to overcome the random effects of sampling error. This can be illustrated with a simple example.

Suppose one is testing whether one color of an animal may be favored over another. How many animals need to be sampled to detect a deviation from some expected ratio? Let's suppose it is a plentiful species, say a fish that we can net and count. Probability theory tells us that the more individuals we count, the closer we come to knowing the actual number of each type of fish in a population. How many do we need in order to be 95 percent confident in our sample (that is, 95 times out of 100 we will be correct within a given range)? As the table below shows, the margin of error decreases as the sample size increases.

Sample Size	Margin of Error (Percent)
100	±9.8
400	±4.9
1000	±3.1
10,000	±1.0

If we sample just 100 fish our estimate may be off as much as 10 percent. One cannot detect subtle selection with such large error. (Poll takers face the same problem of small sample size, which is why they sometimes make errors in forecasting election results).

The challenge of detecting slight selective differences in the wild means that we mostly know of cases where selection is very strong and therefore very fast. The most widely known case is that of melanism in peppered moths. With the onset of the Industrial Revolution, pollution in areas of England and North America altered the coloration and lichen growth on trees where the peppered moth rests. There was a dramatic, rapid increase in the frequency of dark, melanic forms of the moth in industrial areas and a dramatic decrease in the light form. In just fifty years, from around 1848 to 1896, the dark form arose and evolved to as high as a 98 percent frequency in some areas. Haldane estimated from surveys of moth types taken over this period that the selection coefficient against light moths on dark trees was on the order of negative 0.2. A 20 percent disadvantage may seem modest, but, when compounded season after season, it will reduce the frequency of a population very, very quickly. In the last half century, with the enactment of clean air laws, the selection pressure reversed direction and there is excellent documentation of the rapid decline in the dark melanic form from above 90 percent of peppered moths to less than 10 percent in some locations (figure 2.2).

The agents of natural selection on the peppered moth are birds, and here is another variable that makes natural selection difficult to investigate in the wild. Not only do we require sufficient sample numbers of moths, if possible, we want to know about the selective agents acting on them. This can get complicated if there are multiple predators, conditions that vary with habitat and time of day, etc. In the peppered moth case, the rapid rises and declines of the two forms, which occurred in parallel on two continents and were associated with changing industrial practices, were no doubt due to natural selection on color forms.

The peppered moth story is just one example. The natural selection of animal coloration has been studied in land snails, ladybird beetles, desert mice, and other species with definite or probable selective

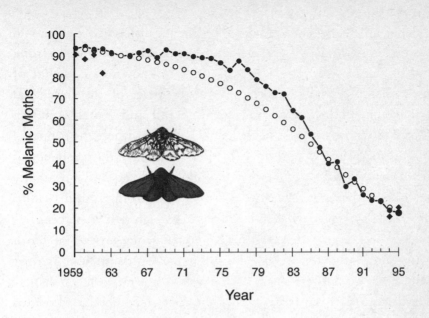

FIG. 2.2. **The decline of melanic moths with the improvement of air quality.** As the selective conditions changed, the frequency of melanic moths has dropped steadily in both the United States (solid diamonds) and Britain (solid circles). The open circles represent the expected decline under a selection coefficient of −0.15. *Figure modified from B. S. Grant et al. (1996),* Journal of Heredity *87:3551.*

agents identified. In some of these species the selection coefficients for certain color types are fairly large (from 0.01 to 0.5).

It is important to appreciate that the long-term studies necessary to study natural selection in the wild require a combination of often heroic personal commitment, continuous financial support, and the even greater fortune of nature's cooperation. Field biology itself selects for human researchers with particularly rare characteristics.

A recent seven-year-long field study of falcon predation on feral pigeons demonstrates the patience and tenacity required for understanding selection in the wild. The pigeons around Davis, California, occur in six different plumage color schemes, and they are a favorite meal of the peregrine falcon. One form of the pigeon is blue-gray with

FIG. 2.3. **Variation in pigeon rump feather coloration.** In urban pigeon populations, some individuals sport white rump feathers (left) that offer some advantage in evading attack by peregrine falcons. *Drawings by Jamie Carroll.*

FIG. 2.4. **The pursuit.** A peregrine falcon hits speeds of 150 mph during its dive. Here, a pigeon in peril. *Photograph by Rob Palmer, courtesy of Alberto Palleroni.*

a white rump of feathers between the base of the tail and lower back; all other forms of the pigeon lack this white rump patch (figure 2.3). A team of researchers, led by Alberto Palleroni of Harvard University, has studied the air war between falcon and pigeon in exceptional detail. Over the seven-year span, they recorded 1485 (!) attacks by five adult falcons on feral pigeons (figure 2.4). The scientists looked for any relationship between the pigeon feather colors and the falcons attack and success rate. To do so, they attached bands to and recorded data on 5235 pigeons—a very large sample.

The researchers found that falcons attacked the white-rumped pigeons less frequently than the other pigeons, and they were generally unsuccessful at catching them when they did try. While the white-rumped pigeons made up 20 percent of all pigeons in the area, they accounted for only 2 percent of all pigeons captured by the falcons.

Palleroni and colleagues attribute this advantage of white rump feathers to their function during an evasive roll that all pigeons execute in midair while under attack. As the falcon dives at speeds over 200 miles per hour, the pigeons dip a wing and roll out of the falcon's path. The biologists suggest that the white patch distracts the falcon as the roll is initiated, providing an extra split second to dodge the attack.

To test their theory further, the researchers *switched* the rump feathers among 756 white-rumped and blue-barred pigeons (by clipping and splicing the feathers with latex glue), released the birds, and recorded their fate during falcon attacks. The capture rates were reversed. The formerly white-rumped birds were captured at the same rate as the rest of the pigeons forms, and the blue-barred pigeons with their transplanted white rumps were protected.

Over the course of this amazing study, the researchers observed a steady increase in the population of white-rumped pigeons over other types. The pigeons are evolving under natural selection.

THE SPECTRUM of studies of natural selection in the wild is expanding to more complex traits than animal color. The evolution of the

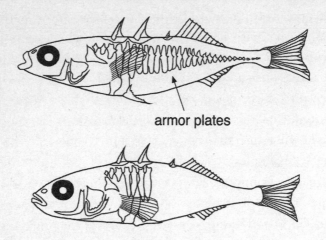

armor plates

FIG. 2.5. **Rapid evolution of armor plating in stickleback fish.** In lakes throughout northern North America, oceanic stickleback fish have adapted to new conditions by reducing the number and size of armor plates, a part of their body skeleton. *Figure from M. A. Bell et al. (2004),* Evolution *58:814.*

threespine stickleback fish is a great example. We know from the geologic record that as glaciers receded at the end of the last ice age, oceanic sticklebacks that invaded lakes and streams became isolated. These freshwater populations then diverged from their oceanic ancestors. The ocean form of sticklebacks typically has a continuous row of more than 30 armor plates running from head to tail. In many freshwater populations, this number has been reduced to a range of from 0 to 9 plates. The selective advantage of plate reduction in lakes and streams may be due to greater body flexibility and maneuverability while swimming.

The evolution of sticklebacks has also been observed in real time. In Loberg Lake, Alaska, oceanic sticklebacks colonized the lake after a chemical eradication program exterminated the resident populations in 1982. Over a span of just twelve years, from 1990 to 2001, regular sampling revealed that the frequency of the oceanic form dropped steadily, from 100 percent to 11 percent, while a form with low plate number rose to 75 percent, with various intermediates making up another small fraction (figure 2.5). The evolution of the Loberg Lake population shows that these fish adapt to freshwater habitats within decades after their arrival from the oceans. Because sticklebacks evolve quickly in new environments, this species is a terrific model for the study of evolution and I will return to them again later, in chapter 8.

These few examples demonstrate the rapidity of evolution and the strength of selection when acting upon variations already present in a population. But where does that variation comes from? And what happens if useful variations are not available? How long might a population have to "wait" for new variations to appear?

The Mutation Lottery: We Are All Mutants

The source of all variety is mutation. This word has many connotations, some of which have contributed to two widespread misconceptions about mutation that I want to dispel right away. The first is that all mutations are bad and therefore must be destructive, not creative. This is entirely false, as the icefish attest. We will see that the odds of evolving a useful mutation are much better than those of buying a winning Powerball ticket. The second notion is that, if mutations are random (which they are), then a random process cannot possibly account for the complexity and order we see in living things. This misconception is based upon a failure to distinguish between mutation and selection. The mutational process is blind, natural selection is not. Mutation generates random variation, selection sorts out the winners and losers. Furthermore, natural selection acts *cumulatively*. Neither Rome nor Romans were built in a day, nor did the Southern Ocean freeze and icefish evolve in an instant. Evolution has shaped icefish, humans, and other species over hundreds of thousands to millions of generations. New mutations are superimposed on, and then incorporated into, an already functioning creature; they do not and need not generate complex functions in a single bound.

In order to appreciate the creative power of mutation, we have to know what kinds of mutations are possible and the frequency of what actually happens in DNA. Fifty years of study have given us a clear picture of the dynamics of DNA. I will provide a brief explanation here of the variety of ways that DNA is altered by mutation.

In order to reproduce, organisms must make copies of their DNA. The copying of DNA is a complex biochemical process. Mistakes happen, and when they are not immediately or correctly repaired, mutations are born. There are many different kinds of mutations. If we think of DNA as being like a written text, then the categories of mutations are just like the familiar kinds of word processing errors. The DNA of a given species ranges from millions to billions of permutations of the four letters A, C, G, and T. The most common mistake is the substitution of an incorrect letter—a typo. But there are many other kinds of events that also occur, such as deletions and insertions of blocks of letters. Copy and paste errors also occur; these result in duplications of text. Groups from just a few letters on up to entire genes, or large blocks of genes, are duplicated at a significant frequency (duplications can thus expand the information content of DNA without causing any difficulties—and we will see in chapter 4 how they are an important source of new functions). Blocks of DNA letters are also rearranged—by inversions and the breakage and joining of parts of text. As a result, in every new individual, there are some new mutations.

The rate of mutation has been carefully studied in many species. In humans, there are an estimated 175 new mutations among the 7 billion DNA letters in every individual. As biologist Armand Leroi has said, "We are all mutants."

Wait! How can this be, you might ask. Aren't mutations *bad*? Sure, some are, but not all of them. We are all generally just fine because these 175 mutations either (1) occur in regions of our DNA that are empty of any meaningful information (2) fall in or near a gene and do not change how that gene works, (3) are compensated for by our carrying two separate copies of most genes, or, (4) affect a gene in such a way that produces an effect within a tolerable range of variation. These variations in size, shape, color, and other physical and chemical properties are what make each of us unique. This is the raw material for evolution.

I'll use a concrete example for which there is plenty of available data to illustrate that the odds of adaptively useful mutations arising are very much on the side of nature (in later chapters I will describe many more examples of adaptive changes). Let's consider a population of wild mice in which every individual is light-colored and lives on sandy soil in a relatively stable habitat. Then, over centuries and millennia, geological activity within their habitat leads to some volcanic eruptions and the formation of lava flows. After that lava cools, it forms black rocky outcroppings. The mice are no longer color-matched to their environment; on the dark rocks they are now visible to predators, such as owls. Darker mice would be better color-matched. So what we want to know is:

* How long will it take for a black-causing mutation to arise in a population of light-colored mice?
* How quickly will that mutation spread?

The answer to the first question is a product of the interplay between chance and time; it is solved in the same way one calculates odds in a lottery. The answer to the second question depends on the interplay between selection and time. We have already seen the math for that.

Mutation rates are well studied in mice. At the Jackson Laboratory in Bar Harbor, Maine, they have been breeding mice for many decades and have data on the occurrence of spontaneous mutations based upon the study of *millions* of mice. In terms of individual letters in

DNA, a mutation occurs in about 2 out of every billion sites per mouse (out of 5 billion total sites). There are about 1000 sites in an average gene that can be mutated. When we multiply 1000 sites per gene times 2 mutations per billion sites, we get the result that one mutation occurs in a specific gene in about 500,000 individuals. This shows how accurate DNA copying is, but that it is not perfect.

We also know that there are multiple genes that, when mutated, can darken fur color. For our example, I'll focus on just one gene for which the Jackson Lab has obtained several mutants, and about which we know a fair amount of detailed biology. This gene is called *MC1R* and we know that there are many sites within it that can be mutated to cause mice that carry just one copy of any of these mutations to become black.

To calculate how often a black mutation will arise, I'll make the following assumptions:

Mutation Rate:	2 per 10^9 sites
Number of sites in *MC1R* that can be mutated to make a mouse black:	10
Number of copies of *MC1R* gene:	2

Let's multiply these together: 10 sites per gene × 2 genes per mouse × 2 mutations per 1 billion sites = 40 mutants in 1 billion mice. This tells us that there is about a 1 in 25 million chance of a mouse having a black-causing mutation in the *MC1R* gene.

That number may seem like a long shot, but only until the population size and generation time are factored in. How long it takes for one mutation to arise is also a function of population size and birth rates. Mice are numerous; every generation they produce a lot of babies. For the mice I am discussing, local population sizes are between 10,000 and 100,000 individuals. To figure out how often the black mutation arises, we also have to make an estimate of birth rates.

These mice have two to three litters of two to five babies per year, so a reasonable estimate is that several babies are born to every female in the population in a year—let's say five on average. If we multiply the number of breeding females times the number of babies per generation, we get the number of offspring born per year. Taking the more conservative figure of 10,000 individuals in the population, half female, and multiplying by the conservative number of five offspring per female, we get 25,000 babies born in the population each year. Now, multiply this number of 25,000 babies per year by 1/25,000,000 (the odds of the mutation arising in an individual), and you get 1 black mouse per 1000 years. Thus, in 1 million years, a black-causing mutation will occur independently 1000 times. Our group of 10,000 light-colored mice will hit the black mutation lotto every 1000 years. If we use a larger population number, such as 100,000 mice, they will hit it more often—in this case, every 100 years. For comparison, if you bought 10,000 lottery tickets a year, you'd win the Powerball once every 7500 years.

Now let's turn to the question of the spread of the black fur color mutation. Once the mutation exists, its spread is a matter of selection and time. The key factors are the selective advantage of the black fur trait, and the effective population size (this is not the total number of individuals, but takes into account the breeding population and other factors; it is designated N_e). The greater the advantage, the faster it will spread, but the greater the size of the population, the longer it will take before every mouse is black. The formula for the average time (t, in generations) required until all of the mice are black is:

$$t = (2/s) \text{ natural log } (2N_e) \text{ generations}$$

Here are some values of t as a function of the advantage s for a population size of 10,000 individuals; note that as the selection coefficient increases, the amount of time for the favorable mutation to spread decreases:

$s = 0.001$	time = 19,807 generations
$s = 0.01$	time = 1981 generations
$s = 0.05$	time = 396 generations
$s = 0.1$	time = 198 generations
$s = 0.2$	time = 99 generations

The estimated selection coefficient for dark mice living on black lava flows is greater than 0.01. Consulting the table above reveals that the dark fur mutation would spread completely in fewer than 2000 generations, or less than 2000 years. This shows that even a mutation of slight benefit will spread in a time span that is short in geological terms.

I have to mention a couple of additional variables here in order to more accurately reflect events and probabilities in nature. One important additional element is the chance that the black mutation will be lost before it has a chance to spread. The mouse or its offspring may not survive to breed, or the offspring in turn may not pass on the black mutation to their offspring. Mutations will be lost from populations as a function of chance and selection. I won't derive the formula here, but the probability that a mutation will spread successfully throughout a large population is approximately 2 times the selection coefficient. In the case above, for a mutation with a small advantage s of 0.01, this would be a 2 percent probability. With an s of 0.05 this would be a 10 percent probability. Still, with 1000 or more occurrences in 1 million years, this means that the mutation could arise and spread 20 to 100 separate times.

I have also not incorporated the effect of migration of animals into the calculations here. Animals do not stay in one place; the mice can and do move back and forth between light sand and the dark lava rock. Furthermore, dark mice are at a disadvantage on light sand, while light mice are at an advantage. This complicates the description of reality, but not the mathematics of the swiftness of evolution.

It is crucial to remember that I have calculated the waiting time for a new variant to appear in an all-sandy mouse population. The reality

FIG. 2.6. **Rock pocket mice.** The melanic form (top) occurs more often on lava flows, the light-colored form (bottom) on sandy backgrounds. *Figure from S. B. Benson (1933), University of California Publications in Zoology 40:1.*

is that in natural populations of mice, pigeons, humans, or most any species, variation usually exists in most traits. For another great example, look at color plate C, which shows eighteen color variations of just one species of garter snake.

I chose the example of mouse fur color because it illustrates how there is plenty of time for mutation and selection in typical populations to produce evolutionary change. But I also chose it because it is a real-world situation. In the Pinacate desert of Arizona, million-year-old black lava flows are inhabited by the rock pocket mouse. In this region, the mouse color forms occurs in two color forms, dark black and sandy-colored. Michael Nachman of the University of Arizona and his colleagues have shown that sandy-colored mice are found most often on sandy-colored habitat and dark-colored mice are found on the black lava rocks (figure 2.6). They have also determined the precise genetic basis of the difference between the dark and sandy mouse fur colors (more on that in chapter 6). The power of this example is that the real-world ecology and genetics demonstrate how natural selection works on a substantial timescale. The important messages from the black and sandy mice, which will be reinforced many times and in many ways in the upcoming chapters, are that mutations can be "creative" and that the main limit to evolution is not so much what is mutationally possible, but what is ecologically necessary.

Time

In 2004, many baseball fans were amazed (and some are still downright euphoric) because something happened that had not occurred in 86 years—my beloved Boston Red Sox won the World Series. Our perspective on time is so skewed to our individual lifetimes. At last, after *86 years*, the fans chanted—can you believe it? It may seem like an eternity, but it is barely a tick of the evolutionary clock.

Even 230 years stretches our imaginations—it was *so long ago* when this country was founded.

A thousand years? The Dark Ages. Unimaginable.

Ten thousand years? That encompasses the entire history of civilization.

My point is that 1 million years is an *immense* amount of time. It is plenty of time for key gene variations to arise, and for some genes to become fossils. It is ample time, many times over, for selection to shape a trait. Our ancestors' brains doubled in size in 1 million years. This is a relatively impressive change of great evolutionary significance, but this time span of change still encompassed at least 50,000 generations. The making of an icefish from a warm-water, red-blooded ancestor spanned 15 to 25 million years—again, plenty of time for a whole suite of changes. Indeed, if anything, the pace of evolution is much *slower* than the maximum possible.

It is crucial to appreciate that selection and mutation operate in nature every day. Every environment impacts continually upon the species that inhabit and reproduce within it. Evolution is an ongoing process. Just as we do not notice the growth of a child or a blade of grass on a day-to-day basis, shifts in climate and the ecological interactions among species are not measurable on a daily scale. But, over longer intervals of time, change is the rule, not the exception. The everyday math of evolution gives us a sense of the interplay and power of chance mutations and natural selection, compounded over time.

It is also crucial to appreciate that selection acts only in the present, within a given environment. It cannot act on what a species no longer needs or uses. And it cannot act on what is not yet needed. Thus, the fittest is a relative, transient status, not an absolute or permanent state.

The DNA Record:
Seeing the Steps of Evolution

If days, years, or careers are often too short a period of time to measure change, how can we more readily observe how the fittest are made? Virtually all of life's history and diversity predates recorded human history. So, how do we find out what happened in the deep past? How can we peer back into the mists of time and determine how species and traits evolved?

The answers to these questions lie in the DNA record.

The steady beat of mutation on the text of DNA has consequences that are vitally important to the study of evolution. The known rate of mutation enables biologists to make predictions about patterns we might see in the DNA record. At the fundamental level of DNA, selection affects the relative success of alternative versions of individual genes. As mutations occur, two or more alternative versions of a gene may exist in a population. The fate of these alternative versions depends upon the conditions of selection. Say there are two versions, A and B; if version A affects survival or reproduction in a superior way to form B, A will be favored. Conversely, if B improves survival or reproduction over A, B will be favored.

There is also a third important possibility, which was not recognized by evolutionary biologists until the sequences of genes and proteins started to be revealed. The third possible fate when two versions of a gene differ is that the difference may be neutral, of no consequence to fitness. Evolutionary biologists once thought that all changes in molecules came about through selection, but the late Motoo Kimura proposed in the 1960s that much molecular change was selectively neutral. The power and importance of Kimura's so-called Neutral Theory is that it provides a baseline assumption about how DNA should vary and change as a function of time, *if no other force intervenes*. When measurements of change deviate from what is expected by neutrality, that is an important signal—a signal that selection has intervened. That signal may reveal that selection has favored some specific change, or that it has consistently rejected others.

Over the course of the next six chapters I will describe how the process of evolution is manifest in the DNA record as species and traits evolve. In the first three chapters, I will show how natural selection rejects changes that are harmful (chapter 3), favors changes that are beneficial (chapter 4), and is blind to changes that are neutral (chapter 5). We will see overwhelming evidence of how selection, or its absence, leaves its signature on DNA. I will begin by focusing on the oldest genes we know, which form part of a continuous DNA record extending back to the earliest cellular life on the planet, more than 3 billion years ago.

Life in the extreme. The West Triplet Geyser in Yellowstone National Park. One of many thermal features containing organisms that thrive at high temperatures. *Photograph by Jamie Carroll.*

Chapter 3

Immortal Genes: Running in Place for Eons

. . . .

To be sure, everything in nature is change but behind the change there is something eternal.

—Johann Wolfgang von Goethe

HE WASN'T LOOKING FOR A NEW KINGDOM.

Microbiologist Tom Brock and his student Hudson Freeze were prowling around the geysers and hot springs of Yellowstone National Park one day late in the summer of 1966. They were interested in finding out what kinds of microbes lived around the pools and were drawn to the orange mats that colored the outflows of several springs.

They collected samples of microbes from Mushroom Spring, a large pool in the Lower Geyser Basin whose source was exactly 163 degrees F, thought at the time to be the upper temperature limit for life. They were able to isolate a new bacterium from this site, a species that thrived in hot water. In fact, its optimal growth temperature was right around that of the hot spring. They dubbed this "thermophilic" creature *Thermus aquaticus*. Brock also noticed some pink filaments around some even hotter springs, which raised his suspicion that life might occur at even higher temperatures.

The next year, Brock tried a new approach to "fishing" for microbes in the hot springs of Yellowstone. His fishing tackle was simple: he tied one or two microscope slides to a piece of string, dropped it in the pool, and tied the other end to a log or a rock (don't try this on your own—you will be arrested and quite likely scalded or worse). Days later, upon retrieving the slides, he could see heavy growth, sometimes so much that the slides had a visible film. Brock was right that organisms were living at higher temperatures than had previously been thought, but he did not imagine that they were living in *boiling water*. And they weren't just tolerating 200 degrees F or more—these organisms were thriving in smoky, acidic, boiling pots such as Sulphur Cauldron, in the Mud Volcano area of the park. Brock's Yellowstone explorations opened eyes and minds to the extraordinary range of life's adaptability, identified bizarre but important new species such as *Sulfolobus* and *Thermoplasma*, and launched the scientific study of what he called "hyperthermophiles," lovers of superheat.

Brock's discovery of hyperthermophiles would, in time, lead to three more discoveries with profound impacts on biology. Brock lumped all of his new species into the classification "bacteria." Under the microscope, they did appear a lot like ordinary bacteria (figure 3.1). But, a decade later, Carl Woese and George Fox at the University of Illinois discovered that various sulfur-, methane-, and salt-loving species actually formed an entire kingdom unto themselves. They were as different from bacteria as bacteria are from eukaryotes (the division of life to which we animals belong, as well as plants, fungi, and protists). This new third domain, or division, of life is now referred to as the Archaea.

The second discovery from Brock's world was a practical one. A heat-stable enzyme that could copy DNA at high temperatures was isolated from *Thermus aquaticus*. This enzyme led to the invention of a new, efficient, and very fast technique for the study of genes in any species. This technique catalyzed a vast expansion in the amount and diversity of DNA information that could be obtained from nature, as

FIG. 3.1. **A sample of microbes from a hot spring.** This scanning electron micrograph reveals the growth of a variety of microbes on a slide immersed in the Obsidian Pool of Yellowstone National Park. *Figure from P. Hogenholtz et al. (1998),* Journal of Bacteriology *180:366.*

well as the creation of a multi-hundred-million-dollar market in DNA diagnostics and forensics.

The third and most recent discovery has emerged from the study of archaean genomes. Scrutiny of archaean genes has revealed critical clues about the making of our own eukaryotic ancestors nearly 2 billion years ago. Still preserved in the DNA of these primitive organisms are many pieces of DNA code that also exist in humans and all other eukaryotes. This shared text forms the remaining traces of an early event that gave rise to the first eukaryote, and is crucial evidence that an archaean was one of our original genetic parents.

In this chapter, we are going to examine some of the oldest DNA text on Earth. The fact that such ancient text has endured over eons of time, against the steady bombardment of mutations that could have erased it many times over, is itself remarkable. But these "immortal" genes are also powerful evidence of two key elements of the evolutionary process—the power of natural selection to preserve the DNA record and the descent of life from common ancestors.

The immortal genes vividly reveal the evidence for one very important but somewhat underappreciated face of natural selection. More thought and attention has been directed to the "creative" dimension of natural selection and how new traits evolve, but this is only one aspect of the evolutionary process. Natural selection also acts to remove, in Darwin's words, "injurious change." I will explain how the effect of the removal of harmful mutations by natural selection is manifest in the DNA records of species, in the form of hundreds of genes that have been preserved across kingdoms of life for more than two billion years. In these immortal genes, the steps of evolution we see are just a matter of "running in place" as the gene's text changes only within narrow limits set by natural selection.

The survival of individual genes over vast geological periods provides more than unimpeachable evidence of the preservative force of natural selection. They are clues to the history of life's evolution from ancient ancestors, a new kind of evidence that Darwin could never have imagined. I will show how these immortal genes are powerful genealogical records that reflect the degree of relatedness among kingdoms and help us retrieve and reconstruct events in the history of life that are not visible in the fossil record.

Looking at DNA and Reading the Code

The skyscrapers of DNA sequences that we now own contain a lot of text, about forty thousand volumes, at a million characters each. The records of some species, such as humans, require a whole encyclopedia of about three hundred volumes while others, such as a bacterium, just a three- or four-volume set. No matter which volume one looks in, the text at first looks pretty much the same, like this:

```
ACGGCTATGGGCTACCAAGGGCTACCAACTACCAAAGTTACGGCTAATCGACAT
AATTGGCTACCAAGACATAACCTGGCTACCAATTACTATGGACGGCCTACGGCG
TCCGCTAATCGACATAACCTTTACTATGGCTACCAAAGTGACATAACCTTTACT
CATAACCTGGCTACCAACCAAGGGCTACCAACTACCAAAATTACTATGGGACAT
TAATCGACATAACCTTTACTAACCTGGCTACCAATTACTATGGACGGCCAATCG
```

etc., for hundreds of pages.

How can such a monotonous text composed of just four different characters encode the instructions for a complex creature? Moreover, how the heck do we read this stuff?

To make sense of the language of DNA, we need to learn how to look at genomes and genes and how to read DNA code. We can then make comparisons between species at many different scales, from very close relatives to vastly different life-forms whose lines split off from one another early in life's history. The clues to evolution emerge from *understanding the meaning of the similarities and differences* we find.

In order to decipher the natural history that resides in the DNA record, we have to have a firm grasp of the language of DNA, and of how DNA information is decoded in making the working parts of living organisms. Don't be intimidated—you can learn the language of DNA. It has a very small alphabet and a very limited vocabulary, and its rules of grammar are simple. The payoff for learning about the DNA code is being able to see, and therefore so much better understand, the process of evolution at its most fundamental level. I understand that new terms can get confusing, so you might want to bookmark this short section for future reference.

Here we go.

Proteins are the molecules that do all of the work in every organism—from carrying oxygen, to building tissue, to copying DNA for the next generation. The DNA of each species carries the specific instructions (in code) necessary for the building of these proteins.

DNA is made of two strands of four distinct *bases*. These chemical building blocks are represented by the single letters A, C, G, and T. The strands of DNA are held together by strong chemical bonds between pairs of bases that lie on opposite strands—A always pairs with T, C always pairs with G—as shown here:

```
~AGTCAGTC~
 |||||||||
~TCAGTCAG~
```

so, if we know the sequence of one strand of DNA, we automatically know the sequence of the other strand. It is the unique order of bases in a sequence of DNA (ACGTTCGATAA, etc.) that forms the unique instructions for building each protein. The most amazing fact about DNA is that all of life's diversity is generated through the permutations of just these four bases. So, if we want to understand diversity, we have to crack the code.

How are proteins built and how do proteins know what their job is? Proteins themselves are made up of building blocks called *amino acids*. Each amino acid is encoded as a combination of three bases or a *triplet* (ACT, GAA, etc.) in the DNA molecule. The chemical properties of these amino acids, when assembled into chains averaging about 400 amino acids in length, determine the unique activity of each protein. The length of DNA that codes for an individual protein is called a *gene*.

The relationship between the DNA code and the unique sequence of each protein is well understood because biologists cracked the genetic code forty years ago. The decoding of DNA in the making of proteins occurs in two steps, which I will now describe. In the first step of decoding DNA, the sequence of bases on one of the strands of the DNA molecule is *transcribed* into a single strand of what is called messenger RNA (mRNA). Then, in the second step, the mRNA is *translated* into the amino acids that build the protein. In the cell, the genetic code is read (from the mRNA transcript) three bases at a time, with one amino acid determined by each triplet of bases (a short example is shown in the right half of figure 3.2).

There are sixty-four different triplet combinations of A, C, G, and T in DNA, but just twenty amino acids. Multiple triplets code for particu-

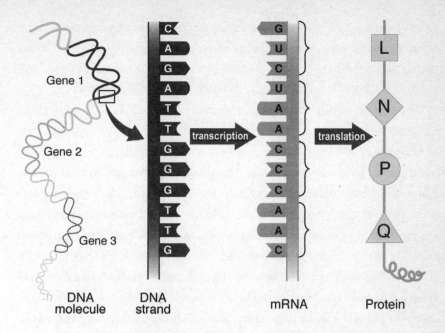

FIG. 3.2. **The expression and decoding of DNA information.** An
overview of the major steps in decoding DNA into a functional protein.
Left, long DNA molecules contain many genes. The decoding of a portion of one gene is shown in two steps. First, the complement of one
DNA strand is transcribed into mRNA. Then, the mRNA is translated
into protein, with three bases of the mRNA encoding each amino acid
of the proteins (shown as the letters L, N, P, and Q here). In mRNA, the
base U is used in place of the T in DNA. *Figure by Leanne Olds.*

lar amino acids (and three triplets code for nothing, and mark the stop-
ping point in the translation of mRNA and the making of a protein—as
periods mark the end of a sentence). Much to our convenience, but also
of profound evolutionary significance, this code is, with few minor
exceptions, the *same* in every species (this is why bacteria can be used to
produce human proteins for pharmaceutical use, such as insulin).

Thus, given a specific DNA sequence, it is easy to decipher the protein sequence that that DNA encodes. However, not every base in DNA
is part of a message for a protein. In fact, a large portion of DNA is
"noncoding." The first challenge biologists have when given long reams

of DNA text is to figure out where the "coded" messages begin and end. With whole genome sequence data this, thankfully, is now carried out largely on computers using algorithms that are really good at searching for, and finding, the needles in the haystacks of DNA sequence.

The coding sequence of an average gene is about 1200 base pairs in length. In some species—particularly, microbes such as bacteria or yeast—genes are very closely packed with relatively small spaces of noncoding DNA between the thousands of genes in the entire genome. In humans, and many other complex species, genes occupy only a small fraction of all of the DNA, and are separated by long intervals of noncoding DNA. Some of this noncoding DNA functions in the control of how genes are used, but a lot of it is what is called "junk." This junk accumulates by various mechanisms and often contains long repetitive tracts with no informational content; it is not purged unless it has adverse effects. I will generally ignore this junk, but it is worth mentioning in order to have a picture of the structure of our genomes as archipelagoes of islands (genes) separated by vast areas of open sea (junk DNA).

The Fates of Genes: The Immortal Core

When scientists look at entire genomes, their first aim is to locate all of the genes within the entire DNA sequence. This allows them to take an inventory of a species' genes that includes the total number of genes and a list of every individual gene. Because biologists have been studying the genes and proteins from species for a while now, we can sort genes and the proteins they encode into categories based upon their function and resemblance to existing genes and proteins.

The most interesting fact from the comparison of genomes is that while the number and kinds of genes differ considerably both between and within the three major divisions of life, great increases in complexity do not require proportionate changes in gene number. As shown in table 3.1, most bacteria possess on average around 3000

Table 3.1. The number of genes in genomes

Bacteria

Aquifex aeolicus	1560
Neisseria meningitidis	2079
Vibrio cholerae	3463
Staphylococcus aureus	2625
Escherichia coli K12	4279
Salmonella typhi	4553

Archaea

Sulfolobus solfataricus	2977
Methanocaldococcus jannaschi	1758
Halobacterium sp.	2622

Eukaryotes

Saccharomyces cerevisiae (yeast)	6338
Drosophila melanogaster (fruit fly)	13,468
Caenorhabditis elegans (nematode worm)	20,275
Tetraodon nigroviridis (fish)	20–25,000
Mus musculus (mouse)	20–25,000
Homo sapiens (human)	20–25,000
Arabidopsis thaliana (plant)	25,749

genes, with the smallest genome of a free-living species containing about 1600 genes. Any two bacterial species may differ in size, however, by as many as 3000 genes. Animals possess roughly 13,000 to 25,000 genes, with some animals differing by many thousands of genes. Note that complex creatures, such as a fruit fly, have only roughly twice as many genes as a single-celled brewer's yeast, and that humans have almost twice as many genes as a fruit fly. But we humans have just about the same number of genes as a mouse.

However, gene number is just a raw figure. More detailed clues about evolution emerge from the direct comparison of the fates of individual genes. The differences in gene number tell us that certain genes must be present in some species and absent from others. Before I discuss some specific comparisons, it is important to think about what we might find when we compare the genes of species that belong to different groups. How similar or different should we expect the genes of different species to be?

Before DNA sequencing was possible, some of the great minds of evolutionary biology in the mid-twentieth century contemplated this question. They knew a bit about mutation and concluded that, over geologic time, mutation would eventually change just about every base pair in a genome. For example, with a mutation rate of about 1 mutation per 100 million base pairs per generation, in 100 million generations, most sites in a gene would be mutated at least once, on average. Given the very short generation times of microbes (on the order of hours), and the modest generation times of plants and small animals (a year or less), then one might expect little trace of similarity between the gene of any two species whose lineages diverged 100 million years ago. Indeed, in his 1963 book *Animal Species and Evolution*, the great biologist Ernst Mayr remarked, "Much that has been learned about gene physiology makes it evident that the search for homologous genes [the same gene in different species] is quite futile except in very close relatives."

But, when we compare different kinds of bacteria with one another, or different animals (whose ancestors diverged well over 100 million years ago) with each other, we find extensive similarities in their genes. For example, when the genome of the infamous delicacy the puffer fish is compared with the genome of the gourmand stupid enough to eat this deadly creature (the human), at least 7350 genes are found that are clearly shared between the two species. Furthermore, the proteins encoded by these genes are on average 61 percent identical. Since the evolutionary lines of fish and other vertebrates (including humans) separated about 450 million years ago, this is a much more extensive similarity than would be expected if mutations were simply allowed to accumulate over time.

More stunning, when we compare the genomes of Archaea, bacteria, fungi, plants, and animals, we find about 500 genes that exist in all domains of life. We know from the fossil record that eukaryotes are at least 1.8 billion years old and the Archaea and bacteria well over 2 billion years old. The genes these organisms all share have withstood more than 2 billion years of the steady bombardment of mutation and stand out as threads of text whose sequence and meaning have not changed significantly despite the vast differences among the species that carry them. These genes are *immortal*.

The functions of immortal genes are central to fundamental, universal processes in the cell, such as the decoding of DNA and RNA and the making of proteins. All forms of life have depended upon these genes since the origin of complex DNA-encoded life early in Earth's history. These genes have survived through an immense arc of time, and life will continue to depend upon this core set of genes as it evolves in the future.

Immortal genes have survived not because they avoid mutation—they are as vulnerable to mutation as all other genes. The genes are immortal in the sense that the gene as a unit endures; however, not every letter of the gene's code endures. This fact can be seen upon more detailed inspection of their DNA sequences and of the sequences of the proteins they encode, and it is a key demonstration of one aspect of the process of natural selection.

Running in Place:
The Conservative Face of Natural Selection

Upon closer examination, we find that while the immortal genes of different species encode very similar proteins, the sequences of bases making up the same genes are less similar than the resulting protein sequences. This discrepancy can exist because of the "redundancy" of the genetic code, which allows different triplets of bases to encode the same amino acid. This feature of the genetic code insulates DNA against injurious changes in that mutations that change a base in DNA do not necessarily change the sequence of the encoded protein. Such changes that do not alter the "meaning" of a triplet are called synonymous changes because the original and mutant triplet are synonyms that encode the same amino acid. Mutations that change the meaning of a triplet and cause one amino acid in the protein to be replaced with another are called nonsynonymous mutations.

It is straightforward to calculate the odds of a mutation being either synonymous or nonsynonymous. There are 64 possible triplets of bases. For any individual triplet there are 3 possible changes at each base, 9 possible single mutations in all. Multiply the 64 triplets by 9 possible mutations and this equals 576 possible different random base mutations. Consultation of the genetic code reveals that 135 of the 576 possible mutations (about 23 percent) are synonymous, while the remaining 77 percent are nonsynonymous. The key prediction from these calculations is that, *without the intervention of natural selection*, the expected ratio of nonsynonymous to synonymous changes in gene sequences is about 3:1 (77:23).

However, in nature the ratio we typically find is about 1:3 in favor of synonymous changes. *This ratio is ten-fold lower than what we expect from random base mutations.* Clearly, only a small fraction of the nonsynonymous mutations that occur are retained over time. What accounts for the far fewer than expected nonsynonymous mutations?

Human	DAPGHRDFIKNMITGTSQADCAVLIV
Tomato	DAPGHRDFIKNMITGTSQADCAVLII
Yeast	DAPGHRDFIKNMITGTSQADCAILII
Archaea	DAPGHRDFVKNMITGASQADAAILVV
Bacteria	DCPGHADYVKNMITGAAQMDGAILVV
"Immortal letters"	D-PGH-D--KNMITG--Q-D---L--

FIG. 3.3. **An immortal gene.** A short portion of the sequence of a protein found in all domains of life (called elongation factor 1-α). Several amino acids, indicated by shading, have not changed over a span of 3 billion years. *Figure by Jamie Carroll.*

Natural selection. There is no other explanation. This skew in the ratio is very clear evidence of a type of selection, called purifying selection, that maintains the "purity" of the amino acid sequences of proteins by ridding them of changes that would compromise their function.

We can see the signature of purifying natural selection in the sequences of the majority of genes, but it is most striking in the immortal genes that have been preserved across all domains of life. For example, many of the proteins involved in the making of key pieces of the machinery for decoding mRNA are shared among all species. Looking at a specific part of just one of these proteins (for simplicity's sake each amino acid is designated by a letter), in representative Archaea, bacteria, plants, fungi, and animals great similarity has been maintained over this part of the protein for over 2 billion years (figure 3.3). Note that fourteen individual amino acids have been absolutely maintained throughout evolution. These fourteen letters (amino acids) are effectively immortal.

If we compare the sequences of the DNA encoding this piece of protein in each species, however, we find that the DNA sequences are less similar than the protein sequence. For example, inspection of the human and tomato versions of this gene reveals that they are identical at just 65 out of 78 positions (83 percent), while the encoded proteins are identical at 25 out of 26 positions (96 percent). The reason for the greater similarity of the protein sequences than of the DNA sequences is the occurrence of 12 synonymous changes in the DNA sequence—these are mutations that are allowed to accumulate.

The pattern of evolution of genes under purifying selection is one of "running in place." That is, the bases may be changing but their translated meaning is not. Consider, for example, the triplet TTA, which encodes the amino acid leucine, in the DNA sequence of a gene. This triplet can change in two different ways and still encode leucine, and these mutated triplets can change further and still encode leucine:

Original triplet	TTA	⟶	leucine
Mutated triplets {	TTG	⟶	leucine
	CTA	⟶	leucine
Double mutated triplets {	CTT	⟶	leucine
	CTC	⟶	leucine
	CTG	⟶	leucine

Most amino acids are encoded by at least two different triplets, and several amino acids are encoded by three or more (or, in the case of leucine, six). So triplets in DNA sequences can "run" (change from sequence to sequence) but selection usually sees to it that they do not run so far as to change the protein's sequence and function.

Selection prevents protein sequences from changing by favoring one particular sequence over variants in which one or more letters of the sequence is altered. If a variant works less well than other forms of proteins, even if that is just 0.001 percent less well, selection over time will, by the algebra we saw in the last chapter, purge the variant from a large population. This purging is so efficient that individual letters in a protein's text can remain unchanged among virtually all species. *Realize that an immortal letter in a protein sequence has experienced mutation again and again, in uncountable numbers of individuals, in millions of species, over billions of years, but that all of these mutations have been purged by selection over and over again.*

From the alignment of protein sequences in figure 3.3, we can see that constraints exist on many amino acids in this protein in that only a few positions are allowed to differ between species. Many more synonymous mutations are allowed than are nonsynonymous mutations. I have shown this relationship for just one gene, but I could have selected *thousands* more genes, either from the 500 or so immortal genes or the majority of other genes in any group of species. This pattern of the strong preservation of the protein sequence at most sites, with the synonymous evolution of the corresponding DNA sequence, and diversity limited to a few sites in the protein, is the predominant pattern of evolution in the DNA record.

DNA sequences that encode the same protein sequence but that are substantially different are unmistakable evidence of natural selection allowing mutations that do not change protein function, while acting to eliminate mutations that would. The preservation of genes among different species over vast periods of time is thus definitive proof of one face of natural selection—its power, in Darwin's words, to "rigidly destroy injurious variations."

But the evolving genomes of species provide more than signatures of natural selection. In the DNA record, there is more information than just the history of a particular gene—there is information about the species that carries it, and about all of the preceding species that also carried it, right back through eons of life's history. Because of the power of natural selection to preserve information that would otherwise be erased in time, genomes contain a record of the history of life.

The new wealth of data from genomes offers unique insights into the deep past that could not be deciphered from any other source. I will close this chapter with the story of the evolution of the domain we belong to (eukaryotes), and the unique contributions that archaea and bacteria appear to have made to our ancestry.

The Making of Eukaryotes:
A Marriage of Two Very Different Parents?

> The time will come, I believe, though I shall not live to see it, when we shall have fairly true genealogical trees of each great Kingdom of Nature.
>
> —Charles Darwin, letter to T. H. Huxley,
> September 26, 1857

Our understanding of the structure of nature has come a long way since Darwin's time. In his day, the living world was divided into just the plant and animal kingdoms. This dual system had been recognized since Aristotle and was formalized by Carl von Linné in 1735. Ernst Haeckel, in 1866, with his remarkable studies of protists, added a third kingdom to life's tree. The bacteria and fungi were not added as full-fledged kingdoms until the twentieth century.

Within this five-kingdom scheme, a higher primary division was also recognized, based upon fundamental differences in the types of cells found in different kingdoms. In 1938, French biologist Edouard Chatton proposed the names "prokaryote" and "eukaryote" based solely on the absence and presence, respectively, of nuclei. These two "superkingdoms" encompassed all of the known living world, until Carl Woese started studying the genes of the kinds of species Tom Brock found in Yellowstone.

Woese believed that bacterial taxonomy was a mess and needed more objective means of determining the evolutionary relationships among species than their appearance or physiological characteristics. He turned to molecules. The possibilities for building species trees based upon DNA, RNA, and protein sequences were quickly recognized by scientists (such as Francis Crick, Emile Zuckerkandl, and Linus Pauling), as soon as protein sequencing began to reveal the similarities and differences in proteins shared among groups of species. The basic idea is quite straightforward. Sites in the text of DNA, RNA, or protein sequences that differ among a group of species, but are shared among subsets of these species, reflect their degree of kinship. Just as we build family trees based upon degrees of genetic relationships, we build species trees based upon their genetic kinship. But, as I will explain, sometimes there is a marriage that really confuses the family tree.

Woese used an abundant type of RNA molecule to make trees of bacteria. But when he included the thermophilic, methane-producing species with conventional bacteria, he found that "these 'bacteria' appear to be no more related to typical bacteria than they are to eukaryotic cytoplasms." He proposed there existed a third superkingdom which, because of the adaptation of these species to the sorts of extreme environments that were presumed to exist early in Earth's history, might be the original or *ur*-kingdom, so he suggested calling this new superkingdom "archaeabacteria." This name was modified later to Archaea, in part to avoid confusion with the bacteria, and the superkingdom category was renamed "domain."

While the division of life into three domains—Eukarya, Archaea, and Bacteria—has held up, the relationships between these three groups has been challenging to sort out. Darwin described the genealogy of species as trees, with speciation producing ramifying branches. But in the world of microbes, unknown to Darwin, some events happen that violate the pattern of treelike evolution. Microbes exchange genes, and some microbes live within other host species in a process called endosymbiosis. These processes enable the transfer of genes between very distant relatives, and thus confuse the family tree. In order to figure out the relationships between eukaryotes, archaea, and bacteria, biologists have to sort out the history of lots of genes, not all of which may have the same family resemblances.

For example, some of the first studies of archaean molecules turned up some striking resemblances between some archaea and eukaryotes. Proteins that archaea use to package their DNA in chromosomes, to transcribe DNA, and to decode messages bear such similarity to those in eukaryotes that it suggested to many that eukaryotes evolved from some archaean. Some of these provocative similarities are in short "signature" sequences in proteins that are shared among some archaeans and eukaryotes, and no other group. For example, there is a short insertion of eleven amino acids in one of the immortal proteins involved in decoding messages. Table 3.2 shows the sequence of this insertion in different eukaryotes and archaea:

Table 3.2. Insert sequences

Eukaryotes	
Human	GEFEAGISKNG
Yeast	GEFEABISKDG
Tomato	GEFEAGISKDG
Archaea	
Sulfolobus	GEYEAGMSAEG
Pyrodictum	GEFEAGMSAEG
Acidionus	GEFEAGMSEEG
Bacteria	(Absent)

The existence of this sequence in two domains but not in the third would be most logically explained by the archaea and eukaryotes being more closely related to each other on life's tree than to bacteria. The resulting picture of the tree would posit that there was a common ancestor of all three domains (the last "universal" common ancestor, or LUCA) that then split into two domains, the Bacteria and Archaea,

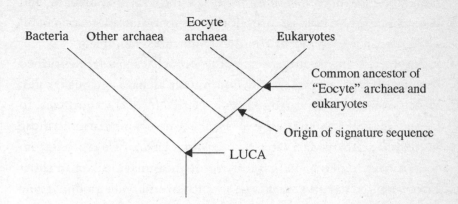

FIG. 3.4. **A "conventional" tree of life.** The tree depicts all domains arising by the splitting of lineages. *Figure by Jamie Carroll.*

and the eukaryotes arose later from a branch of the Archaea. The tree of life would then be as shown in figure 3.4.

However, sequencing of whole archaean and bacterial genomes revealed, somewhat unexpectedly, that the majority of archaean genes show the greatest similarity to bacterial counterparts. Then, as more eukaryotic genomes were sequenced, their analysis suggested that many eukaryotic genes were more related to those of bacteria than of archaea. The story was taking on the nature of one of those riddles like "If your sister is also your aunt, then who is your father?" In short, the answer to the question of which groups are most closely related was muddied.

The resolution of the riddle stems from the pursuit of a key observation. It was noted that most of the similarities between archaea and eukaryotes were in so-called informational genes whose products dealt with the copying and decoding of DNA. Furthermore, most of the similarities between eukaryotes and bacteria were in operational genes involved in the metabolism of various nutrients and basic cellular materials. From the viewpoint of eukaryotes, it appeared as though they got their "brains" (informational genes) from one parent, and their "looks" (operational genes) from another.

This raised the specter that eukaryotes were the product of a mixed marriage—a genetic fusion of archaean and bacterial parents. The notion of a fusion between vastly different species is not new. In 1970, Lynn Margulis proposed that mitochondria and chloroplasts, two key energy-producing organelles in eukaryote cells, arose from bacteria living within eukaryotes (this fusion process is endosymbiosis). This view is now widely accepted.

But what about the making of a eukaryote from an archaean and bacterial ancestor? Maria Rivera and James Lake of UCLA have concluded that, indeed, eukaryotes are of dual origin, from parents belonging to different branches of life. Rivera and Lake analyzed bacterial, archaean, and eukaryotic genomes for the sets of genes shared in all, all but one, all but two, all but three, etc., of the major divisions within these three domains. Comprehensive analysis of these patterns of shared genes indicate that the eukaryote genome is the product of a fusion between a relative of a type of archaean and a type of bacterium. Because symbiotic relationships are common among organisms living together (for instance, Yellowstone's *Thermus aquaticus* gets its energy from photosynthetic cyanobacteria that color those bacterial mats), and this occasionally leads to endosymbiosis, the likely explanation for eukaryote origins was as a product of the fusion of genomes between an endosymbiont and its host. The resulting base of the tree of life is then not a trunk, but a ring from which our tree ascends and branches (figure 3.5).

So, if you happen to visit magnificent Yellowstone, don't turn away from the stinking, boiling soup of those hot cauldrons or be revolted by the colorful strings and mats of slime that ooze around their edges. That's no way to respect one's relatives, no matter how distant. Ponder the amazing fact that you share hundreds of genes with members of this community. And, in this sort of community, somewhere an unfathomably long time ago, perhaps along a deep sea vent, in a belch of methane there emerged the ancestor of all of the familiar and visible kingdoms on Earth.

Of course, if all natural selection did thereafter was to maintain the status quo within very strict limits, life would be uniform and

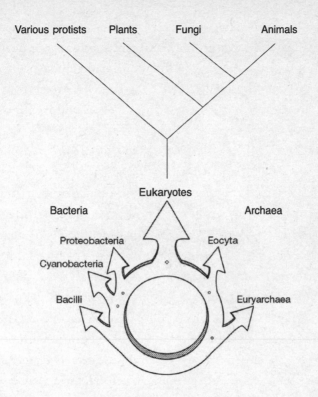

FIG. 3.5. **A new picture of the tree of eukaryotes.** The DNA record indicates that an ancient fusion of some type of archaea with some type of bacteria contributed to the origin of eukaryotes. The base of the tree is a ring, not a conventional trunk. *Adapted from M. Rivera and J. Lake (2004)*, Nature *431:152.*

unchanging, and not the riot of diversity we see in the world today and in the past 3 billion years of the fossil record. The figures on gene number in table 3.1 tell us that vast differences exist in gene content between life-forms. Beyond the core of 500 or so immortal genes, species vary widely in gene number. The differences in gene number tell us that, in the course of evolutionary time, new genes must be born. They are indeed, and that creative dimension of evolution will be the focus of the next chapter. It is also a hint that genes might also die. They do die, and I will take up that twist and what it can tell us about evolution in chapter 5.

Colobus monkey in Kibale Forest, Uganda. This monkey can see the more nutritious leaves and digest them because of two key evolutionary innovations, full color vision and a ruminating stomach. *Photograph by Cagan Sekercioglu.*

Chapter 4

Making the New from the Old

. . . .

Preserve the old, but know the new.

—Chinese proverb

IT IS BREAKFAST TIME IN KIBALE FOREST.

High above the forest floor, a troop of magnificently painted black and white colobus monkeys hurl their way through the canopy toward a fresh meal. Surrounded by the dense foliage of this Ugandan rain forest, it would seem that the menu choices are limitless. But the colobus pass by the abundant greenery and set their eyes on plants bearing reddish leaves. The only monkey that lacks thumbs (their name comes from the Greek *kolobus*, meaning "mutilated" or "docked"), the colobus pull the branches bearing the tender young leaves to their mouths with their four agile, long fingers.

Lower down, nearer to the forest floor, a few of the five hundred resident chimpanzees are noisily making their way to some Natal fig trees. They grab a few leaves for chewing, but it is the ripe yellow and red fruits they are after.

Eschewing the immature green fruits, the chimps pluck a few choice samples and find a comfy perch nearby.

The colobus monkeys and the chimpanzees are selecting their breakfasts using a sense that nonprimate mammals lack—full color vision. All apes and Old World (African and Asian) monkeys have trichromatic vision, which allows them to see across the visible spectrum from blue to green to red. Most other mammals are dichromats—they can see blue and yellow hues but cannot perceive or distinguish red from green. Trichromatic vision is important because it is the younger leaves that are most nutritious and tender, and easier to digest. In the tropics, over half of all plant species bear young leaves that are red. This color difference, invisible to other browsers, is exploited by primates to harvest the more nutritious leaves in the forest.

The colobus monkey will digest his leaf breakfast using a capability that neither the chimp nor other apes have, and that only his very closest monkey relatives possess. The colobus is a ruminant. He can specialize on a diet of leaves because he has an unusually large stomach with multiple chambers. A thirty-pound adult will consume four to six pounds of leaves a day, which gives him quite a pronounced potbelly when he sits and rests. Bacteria in the colobus's gut help to digest the large bolus of leaves as it travels slowly through his digestive system, and unique enzymes break down key nutrients that are released from the bacteria.

The visual and digestive systems of these primates raise one of the greatest questions in biology: How do new capabilities arise? In this chapter, we are going to explore the ways through which species acquire new abilities and modify existing talents. The main message of the chapter will be about how new functions and genes are made from "old" genes. I will explain how the accidental, chance duplication of genes provides spare genetic parts for the evolution of new functions, and how both new and old parts are fine-tuned to species lifestyles.

I could have picked many different kinds of examples of the invention and fine-tuning of species capabilities. However, I am going to focus almost entirely on the origin and evolution of color vision, for several very important reason. First, the utility of this sense to its owners is obvious. Second, animals that live in different habitats (ocean, savannah, forest, caves, underground, etc.) have visual systems that are amazingly well adapted to life in these habitats. Third, we understand the biology and physics of color vision very well, so that we understand the large and small differences in species abilities and in the colors that they see. We have learned that there is a wide world of color—the ultraviolet—that humans cannot see, but that birds, insects, and many other animals use to find food, mates, and one another. Fourth, the genes underlying color vision and its evolution have been studied perhaps more intensely than those affecting any other trait. Together, the combination of these elements has produced the deepest body of knowledge linking differences in specific genes to differences in ecology and to the evolution of species.

The body of new knowledge I will discuss in this chapter provides concrete, direct evidence of three elements of the evolutionary process—natural selection, sexual selection, and the descent of species with modification. This trifecta comes from tracing the steps of evolution in DNA. In order to see the steps of evolution, we will identify important differences among particular species, determine when these differences arose, and decipher how specific changes at the DNA level relate to specific abilities in nature. We will track two kinds of information in DNA. First, in order to know when and in what species events occurred, we will track newly discovered, unique landmarks in DNA that give us a clear picture of species relationships. And second, we will scrutinize the DNA code of genes that determine color vision. The signature of natural selection is written in the code of these genes.

Journalist Rex Dalton, remarking on the world that birds see but which is invisible to us, wrote, "If you want to get inside an animal's mind, it helps to see the world through its eyes." I will begin by explaining how animals see colors, and then we'll watch evolution unfold through those very eyes.

Seeing the Rainbow

Humans' vision of the natural world is unique. We see the colors and hues we do because of the fine-tuning of sets of molecules that detect light in the cells of our retinas, and the wiring of these cells to our brains. Other animals have different sets of molecules and/or they are tuned to detect different parts of the rainbow. In order to understand what we or other animals see, we have to know about light and color, the molecules that detect light, and the cells in the eye that gather color images.

Our eyes are very sensitive to what we anthropocentrically call visible light, which is a narrow band of the entire spectrum of electromagnetic radiation. White light is a mixture of the colors of the visible spectrum that ranges from violet to blue, green, yellow, orange, and red. These colors have different wavelengths ranging from about 400 nanometers (nm) (violet) to about 700 nm (red) (figure 4.1). The colors of objects are due to the wavelengths of light that are absorbed or reflected, and this depends upon their molecular composition. Grass is green, for example, because it absorbs all wavelengths of light except green; the green light, whose wavelength is about 520 nm, is reflected. The sky appears blue to us because the shorter wavelengths of sunlight, most of which are blue, are scattered by the atmosphere. The remaining light appears yellowish because this is the color of white light minus the blue. Sunsets appear orange because when the sun nears the horizon, light travels a longer distance through the atmosphere before it reaches our eyes, and more of the blue light is scattered, leaving the orange color. Sunlight also contains shorter wavelengths of light that we cannot see, such as *ultraviolet radiation* that is scattered by the ozone in the atmosphere. That which still gets through and is absorbed by our skin causes sunburns in pale folks like me, tanning in those more fortunate. Longer wavelength light such as the heat produced by fires, is *infrared radiation*. Again, this can't be seen by the

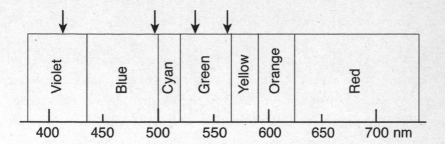

FIG. 4.1. **The color spectrum and wavelengths of light.** The visible violet, blue, green, and red hues are produced by lights of different wavelengths (in nanometers, nm). Human color vision is tuned to four different wavelengths (arrows) through four different opsin proteins. *Drawing by Leanne Olds.*

naked eye, but it can with the aid of night vision goggles, devices with infrared detectors that can see warm objects.

Color vision begins when light of a particular wavelength strikes the visual pigments in our retina. These visual pigments are made up of a protein, called an opsin, and a small molecule called a chromophore, which in humans is a derivative of vitamin A. The light sensitivity of a visual pigment is determined by the exact sequence of the opsin protein and how the chromophore interacts with it. This interaction results in fine *spectral tuning* so that each visual pigment is tuned to a particular wavelength of light. Humans have three different visual pigments, which are sensitive to short, medium, and longer wavelengths of light; these are referred to as SWS, MWS, and LWS opsins, respectively. The three human opsins are tuned to 417 nm (SWS, blue), 530 nm (MWS, green), and 560 nm (LWS, red), respectively, and together provide our color vision. A fourth pigment, *rhodopsin* (497 nm), is used primarily to see in dim light. Light with wavelengths less than 400 nm (ultraviolet or UV) or greater than 700 nm (infrared) is invisible to us, but as I will explain later, many animals see in the UV range.

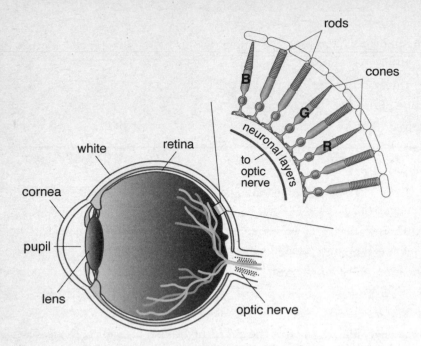

FIG. 4.2. **How color is detected in the retina of the eye.** Light enters the eye through the lens and then strikes the retina. Photoreceptor cells in the retina are of two types, rods that detect dim light and do not produce color images, and cones that contain one of three opsin types in humans that are tuned to red (R), green (G), or blue (B) light wavelengths. Stimulation of the rods or cones is integrated by neurons that feed into the optic nerve. In turn, the optic nerve leads to visual centers of the brain. *Drawing by Leanne Olds.*

In our eyes, we have two distinct kinds of *photoreceptors,* rods and cones, named for the shape of their outer regions, which are packed with visual pigments (figure 4.2). In general, rods are very light-sensitive, and most useful at night and under low light, but they are not capable of discriminating wavelengths, so we are color-blind at night. Cones are most useful in brighter light, and they form the system for everyday color vision.

When light hits the chromophore, it causes a series of very rapid changes in the visual pigment. In a matter of a thousandth of a second, the pigment is "excited" and this excited state causes a "firing" of the photoreceptor cell. The inputs of our retinal photoreceptors eventually reach and are integrated by visual areas in the cortex of the brain. For an object to be perceived as colored, at least two kinds of cone photoreceptors must be triggered. The specific color that is perceived by the brain is determined by the relative level of excitation of each kind of cone. If only one type of cone is present, objects are perceived in shades of gray.

Each of the opsins is encoded by a separate gene. The three opsin genes of humans (SWS, MWS, and LWS) are also present in chimpanzees and other apes. However, most other mammals have just two opsins and genes, while birds and fish have four or more. So, clearly the number of opsin genes has changed in the evolution of animals. The evolutionary history of our opsin genes is an example of *gene duplication*, which is one important way in which information is increased in DNA. In this process, an existing gene is duplicated, and then the "new" and "old" gene go their separate ways, evolving into distinct genes with separate functions.

While this has definitely happened to opsin genes in the course of vertebrate evolution, we want to know more—namely, in which animals were opsin genes gained (or lost)? It is not sufficient merely to say that two species differ—to see the steps of evolution we need to develop a kind of before and after picture. In order to figure out the true history of opsin genes, we need to understand the exact relationships between different animal groups. Accurate knowledge of species' relationships allows us to know the direction of the evolution of a trait and to deduce the status of traits and genes in common ancestors. If, for example, two related species share a trait or gene, the most likely explanation for that condition is not mere coincidence, but that their last common ancestor also possessed that trait or gene. Deciding when the origin of a trait occurred depends upon knowing species' genealogies. To see how the genealogies of species relevant to the origin of our color vision were determined, it's necessary to have a brief understanding of how species' relationships are now determined in the DNA era. I will highlight the powerful, newly discovered land-marks in DNA that can be used to resolve species histories with unprecedented clarity and confidence.

From Molecules to Trees

Ever since Darwin recognized the treelike branching pattern of evolution, biologists have been building trees to represent the relationships of living and fossil species. For the larger part of biology's history, trees were constructed based upon the outward appearance of living species, as well as those of species in the fossil record. The resemblances or lack of resemblances were, however, often discovered to be misleading, or at least a source of contention and disagreement among different biologists such that many, many different trees have been drawn of virtually all major groups.

Beginning a few decades ago, the sequences of molecules and their degree of resemblance were used to trace species relationships. Because genes are inherited, the sequences of genes and the proteins they encode are passed to descendants of species. Changes that occur in one species are passed to its descendants and so on throughout life's long genealogy. Therefore, the degree of similarity in DNA is an index of the relatedness of species.

When biologists want to examine DNA or protein sequences for clues to genealogy, they have potentially *thousands* of genes from which to choose. There are some criteria we apply, but typically (for practical reasons, such as cost) just one or a few genes are selected out of many possible candidates. Once the sequences of the genes of interest are collected, there are many sophisticated mathematical and statistical formulas that are used to figure out the species tree. The key principle is that these formulas seek to identify the tree that best fits the data. While it may seem a minor point, it is important to know that because gene sequences are composed of just the four simple characters A, C, G, and T, the signal in the data as to which species are most related can be uncertain or confused if there is not enough data. I do not want to give the impression that every gene will give the same, correct, or clear answer. So, one way that potential trees are tested further is by examining whether the same trees are obtained with different sequences, and larger amounts of sequences.

Fortunately, there is, however, an altogether new way of deciphering species' relationships. It also relies on DNA, but rather than being based on the degree of sequence similarity, it looks for the presence and absence of certain landmarks in specific places in species DNA. These landmarks are produced by accidental insertions of junk DNA sequences near genes. Particular chunks of junk DNA, called long interspersed elements (LINES) and short interspersed elements (SINES), are very easy to detect. Once a SINE or LINE is inserted, there is no active mechanism for removing it. The insertion of these elements marks a gene in a species, and is then inherited by all species descended from it. They are really perfect tracers of genealogy. These insertion events are very rare; therefore, their presence in the same place in the DNA of two species can be explained only by the species sharing a common ancestor. The inheritance of variable markers in DNA is the same principle applied to paternity testing in humans. By surveying the distribution of a number of elements that arose at different times in different ancestors, biologists have sufficient forensic evidence to determine species' kinship beyond any doubt.

But let's not be satisfied with mere description of how genealogies are deciphered. Throughout this book, I want to develop an appreciation and understanding of the nature and quality of evidence we have. Seeing is the key to believing, and understanding. So let's see how a very important tree, that of humans and other primates, was resolved using SINES as landmarks. We'll then use this tree to unravel the origin of color vision.

The first step in making the tree is to identify a set of SINES to be surveyed. Given the human genome sequence, this is straightforward and there are thousands of SINES in human DNA to choose from. Then, to survey other primates, specific regions of their DNA are

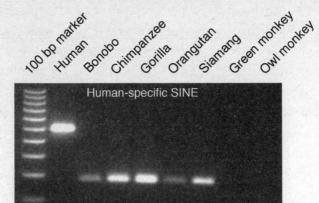

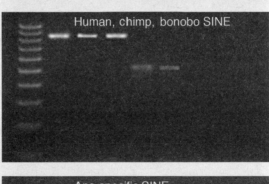

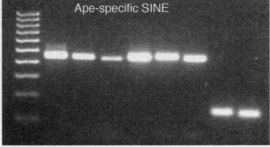

FIG. 4.3. **DNA typing and hominoid evolution.** The presence of particular SINES in the DNA of different species is indicated by larger DNA bands in some lanes, relative to other species. The sharing of SINES indicates the evolutionary relationships among species. *Figure from Salem et al. (2003),* Proceedings of the National Academy of Science *100:12787; copyright © 2003 by the National Academy of Sciences, U.S.A.*

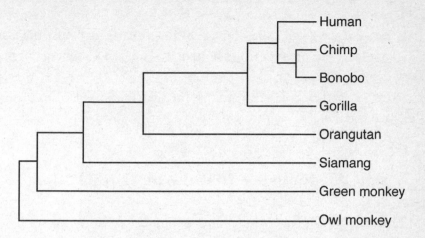

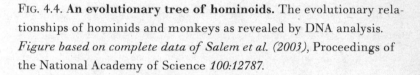

FIG. 4.4. **An evolutionary tree of hominoids.** The evolutionary relationships of hominids and monkeys as revealed by DNA analysis.
Figure based on complete data of Salem et al. (2003), Proceedings of
the National Academy of Science *100:12787.*

examined. Most SINES are about 300 base pairs long, so if another
species' DNA contains the same SINE as that in the human, the region
containing the SINE is about 300 base pairs longer than if the region
lacks the SINE. In figure 4.3, I show several actual sets of data from a
study by Abdel-Halim Salem and colleagues at Louisiana State
University and the University of Utah. The presence or absence of a
SINE is determined visually by the relative position of bands of DNA
in a thick gel used to separate DNA of different sizes. In figure 4.3,
one SINE is human-specific; the second is shared by humans, bonobos, and chimps; and a third is shared by humans, bonobos, chimps,
gorillas, orangutans, and siamangs. Analysis of more than 100 SINES
revealed sets that were shared among subsets of primates including all
six apes examined, five apes, four apes, etc., as well as SINES that were
shared only between bonobos and chimps, and that were unique to
humans, chimps, or bonobos. None of the SINES was found in an owl
monkey. The number of unique shared SINES is evidence of kinship
and, when plotted, yields the tree shown in figure 4.4. The tree depicts

chimpanzees as our closest relatives, bonobos as the chimps' closest relative, and the gorilla and orangutan lines diverging before the last common ancestor of chimps and humans. There is no doubt about the accuracy of this tree.

Now, given this tree, let's consider the distribution of color vision and opsin genes.

The Meandering History of Color Vision

All of the Old World (African and Asian) apes and monkeys have trichromatic color vision and three cone opsin genes, while the New World monkeys, as well as rodents and other mammals, generally have dichromatic vision and two opsin genes. Given the tree in figure 4.4, we can deduce that full color vision arose in an ancestor of the Old World primates, after the separation of the Old World and New World lineages. Furthermore, because the Old World primates possess a third cone pigment, this opsin gene must have arisen after this split as well. This tells us that human color vision dates from a deep, Old World ancestor and was not invented independently during the recent course of hominid evolution.

The existence of two cone opsins in other mammals—squirrels, cats, dogs, etc.—suggests that the presence of two cone opsins and dichromatic vision was the condition in a common ancestor of mammals. But before we leap to the conclusion that full color vision in primates is a unique "advance," we have to consider the vision status of other vertebrates besides mammals. And here's the rub. Birds have fabulous color vision; so do reptiles and many fish, such as the goldfish. Members of these groups have at least four opsin genes. Indeed, in more primitive groups of vertebrates, such as the jawless lamprey, we find *five* opsin genes; this indicates the presence of color vision very early in vertebrate evolution, before the split between the lines of jawless and jawed vertebrates. So, when considered in the context of the whole evolutionary tree of vertebrates, the nonprimate mammals are impoverished with respect to color vision and opsin genes. From the distribution of color vision and opsin genes throughout vertebrates, we can deduce that the pattern of opsin gene evolution in our history was one of initial abundance, then a loss in the ancestors of mammals, and then expansion again in an ancestor of Old World primates.

You might be wondering: If color vision is so great, why would it ever be lost? The most likely explanation has to do with the evolution of nocturnality in mammals. Early mammals were small and lived a cryptic, nocturnal lifestyle in ecosystems dominated by bigger animals, such as the dinosaurs. The evolution of nocturnality shifted the dependence of these animals from color vision in bright light to vision in dim light and darkness, and full color vision was lost. (We will see many instances in the next chapter how such shifts lead to loss of genes, including color vision genes.)

We can say for certain when, relative to primate and mammal evolution, our third opsin gene evolved. But there remains the question of how? How did this new gene expand the spectrum of color vision? We can see the steps taken in the evolution of color vision in the code of the opsin gene. So, we will look at the actual code of the opsin protein and how our two red-green opsins are different from each other, and the exact differences that make them sensitive to different colors. The "tuning" of opsins in the adaptation to specific environments is a general phenomenon in color vision. I will first show how our opsins are tuned and the evidence for their adaptiveness in primates. I will then describe some examples of how different species opsins have been tuned to light of other wavelengths as they adapted to different environments and stimuli.

The Red/Green Show

Rats, mice, squirrels, rabbits, goats, and other mammals have one MWS/LWS opsin whose maximal absorbance is at wavelengths from about 510–550 nm. This opsin is encoded by a single gene. In contrast, humans have two opsins (one for MWS, one for LWS), encoded by two genes on our X-chromosome that lie together as a head-to-tail tandem pair. These opsins are very similar (98 percent) to each other at the level of their DNA code. Their position as next-door neighbors in our DNA and their great similarity are telltale signs that they arose by the duplication of a single MWS/LWS gene in a primate ancestor. Gene duplications are a common form of change in DNA—many of our genes are members of multicopy families that have expanded in the course of evolution. The expansion of gene number increases the information that selection can act upon, and a common pattern of duplicated genes is for their functions to become different. This is exactly the case with our two X chromosome opsins.

Our pair of opsins and those of other trichromatic primates are most stimulated by light with wavelengths of about 530 (green) and 560 (red) nm (these points of greatest stimulation are referred to as absorption maxima). Advances in understanding the functional properties of opsins have revealed that it is very easy to shift the absorption spectrum of individual opsins by changing particular amino acids. The maintenance of the 530 and 560 nm absorption maxima throughout the trichromatic primates suggests that there is selection pressure to maintain this precise spectral separation.

There are just fifteen amino acid differences between the green and red pigments. Biologists have been able to pinpoint which of these differences are responsible for the different functions of each pigment by making precise replacements of one amino acid with another and measuring the effects of these replacements on the spectral properties of each opsin.

Three sites, at amino acid positions 180, 277, and 285, appear to account for most of the 30 nm difference in the absorption peaks of our green and red pigments. These differences and the shifts in light absorption accounted for by each difference are shown in table 4.1.

Together, the evidence from the duplication and function of our opsin genes indicates that an ancestral MWS/LWS pigment gene was duplicated and that the two duplicate genes diverged from each other,

Table 4.1. Amino acids at positions of human opsins

		180	277	285
Pigment	green	A	F	A
	red	S	Y	T
	shift	3–4 nm	7 nm	14 nm

one being tuned to 530 nm, the other to 560 nm, primarily by changes that occurred at these three sites (figure 4.5).

The red/green visual pigment duplication must have followed the split between Old World and New World primates. This is thought to have occurred about 30 to 40 million years ago, shortly after the geologic split between the African and South American continents. The evolutionary changes at those three amino acid sites following the duplication appears to have imparted a substantial advantage on the species that bore them. Only trichromatic primates now inhabit Africa and Asia. If there were other primates around at the time of the origin of trichromatic vision and they did not possess these visual pigments (which seems likely), they have no living descendants today.

Of course, we were not around 30 to 40 million years ago, so one could say this is merely an informed deduction. But there is other evidence supporting the current importance of color vision in primates. One clue is the frequency of color blindness in the wild. Color blindness is common in humans—up to 8 percent of Caucasian males are color-blind due to abnormalities in their X-linked red/green opsin genes. However, in the wild, color blindness is very rare. A study of 3153 macaque monkeys revealed only 3 color-blind individuals (less than 0.1 percent). Given the high frequency of color blindness in humans (where color vision is certainly under less intense selection now, if at all) and the low frequency of color blindness in the wild macaque, this suggests that selection is maintaining color vision in these monkeys and other trichromatic species.

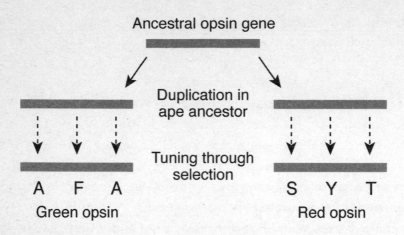

FIG. 4.5. **The duplication and fine-tuning of an ape opsin gene.** In a common ancestor of apes and Old World monkeys, an opsin gene was duplicated. Mutations in the two genes arose that tuned the respective opsin to green and red light wavelengths; these mutations were favored by natural selection. *Figure by Leanne Olds.*

The second source of evidence for the ecological importance of trichromatic vision in primates comes from field observations of food selection by trichromatic and dichromatic primates. Peter Lucas at the University of Hong Kong, Nathaniel Dominy (now at the University of California–Santa Cruz), and their collaborators have undertaken detailed studies of the food preferences and consumption patterns of colobus monkeys and chimpanzees in Uganda, lemurs in Madagascar, and spider monkeys in Costa Rica. They found a consistent preference among trichromatic animals for leaves that were redder, an indicator of high protein levels and low toughness. Most of the primates studied also incorporate fruit in their diet, and there are different preferences among the animals in fruit color as well. However, Lucas and Dominy argue that full color vision is most significant in leaf consumption, particularly when fruit sources are out of season or scarce.

Tuning to red and green, therefore, seems to have its advantages. But red and green are only one part of the color spectrum, important in a plant-dominated forest, for sure. Animals live in many different kinds of environments, including some in which red/green vision is utterly useless, such as the sea.

In the Deep Blue Sea

At greater ocean depths, sunlight is filtered to such an extent that only dim light is available. The detection of dim light is the responsibility of rod photoreceptors, and the opsin in these cells is rhodopsin. Human rhodopsin and the rhodopsins of most terrestrial mammals are tuned to maximally absorb wavelengths of about 500 nm.

At depths of 200 meters or so, only a narrow band of blue light is available, with a wavelength of about 480 nm. Remarkably, the rhodopsins of deep-sea fish and dolphins are "blue-shifted"—that is, they maximally absorb light that is 10 to 20 nm shifted toward the blue end of the light spectrum compared with the rhodopsins of terrestrial mammals. The fine-tuning of rhodopsin has been carefully studied in ocean-dwelling species such as the bottlenose dolphin, common dolphin, pilot whale, Sowerby's beaked whale, John Dory fish, and various eels. The individual amino acids responsible for the spectral differences in rhodopsins have been pinpointed in the laboratory by replacing the individual amino acids or groups of amino acids that are found in one species with those of another. In the bottlenose dolphin, three sites in rhodopsin, at positions 83, 292, and 299, are primarily responsible for the net 10 nm blue shift from the rhodopsin of terrestrial mammals (figure 4.6). The beaked whale rhodopsin is even further blue-shifted (484 nm) and differs from the dolphin rhodopsin at one key site (299).

The evidence that this blue shift in rhodopsin is linked to adaptation to life in deeper waters is supported by analysis of rhodopsins in deep-ocean-dwelling eels and shallow freshwater eels. The eels that live at depths have a blue-shifted rhodopsin that contains the exact same three critical amino acids as the beaked whale. The shallow-

		Amino acids at critical positions of rhodopsin			
	Species	83	292	299	Wavelength of maximum absorption
Deep water	Deep-sea eel	N	S	A	482
	Bottlenose dolphin	N	S	S	489
	Beaked whale	N	S	A	484
Terrestrial or shallow water	Cow	D	A	A	499
	Human	D	A	A	499
	Freshwater eel	D	A	S	502
	Manatee	D	A	S	502
	Harbor seal	D	A	S	501

FIG. 4.6. **The tuning of rhodopsins is associated with the depths at which aquatic animals live.** Deep-water fish and cetaceans have blue-shifted rhodopsins compared with shallow-water and terrestrial animals. The amino acids in three key positions are often the same in different animals that live at similar depths. *Figure by Jamie Carroll.*

water eel has a rhodopsin whose absorption maximum is similar to terrestrial mammals, and that is identical at the three critical sites to rhodopsins of the harbor seal and the manatee, two surface-dwelling, shallow-water mammals with typical rhodopsins.

This correlation between blue-shifted rhodopsins and life in the deep is striking and makes a lot of sense. It is reasonable to suggest that natural selection has tuned rhodopsin to different environments. But there is a more compelling reason to infer the action of natural selection, and that is consideration of the evolutionary relationships of the species listed in figure 4.6. Dolphin and beaked whales are cetaceans, which is a group of mammals that evolved from a terrestrial ancestor that returned to the water. It may be a bit of a surprise, but cetaceans' closest living relatives are hippos, deer, cows, pigs, and camels. We know this from the DNA record of SINES and LINES, and from other DNA sequences. Since the rhodopsins of the terrestrial relatives of cetaceans are tuned to 500 nm, we can confidently conclude that the changes in dolphins and whales occurred after their evolutionary line broke away from that of these other mammals.

But eels are fish, whose lines of evolution split off from other kinds of vertebrates several hundred million years ago. This means that the same precise differences in the deep- and shallow-water eels and terrestrial and marine mammals evolved *independently*. When two species or groups of species evolve the same exact amino acids in a protein in adapting to similar environments, as the cetaceans and deep-water eel have here, this is very strong evidence of natural selection for the same adaptation. (This example of rhodopsin evolution is just a glimpse of many more examples of evolution repeating itself, which will be the entire focus of chapter 6.)

Now let's go beyond red, green, and blue to colors we don't see, and a whole world of animal interactions based on ultraviolet vision.

Beyond the Rainbow

A dozen years after the *On the Origin of Species*, Darwin published *The Descent of Man and Selection in Relation to Sex*. This was Darwin's first detailed treatment of human evolution, but perhaps even more important than that topic was his presentation of a new and fundamental insight into evolution: the importance of the role of the opposite sex in the evolution of traits, which he called "sexual selection." Both sexual selection and Darwin's development of the theory are generally less well-known than his theory of natural selection, but biologists view sexual selection as one of the most important and interesting forces in the evolution of animals—an arena where the assessment of the "fittest" is directly tied to mating success.

Darwin was captivated by bird plumage patterns and gave considerable thought and many pages of description to the gaudy colors and fancy feather patterns of all sorts of species. He was especially interested in how female mating preferences could lead to the evolution of elaborate ornaments in males, such as the peacock's tail. Ever since Darwin, birds have been a favorite subject for the study of sexual selection. But until very recently, such studies were often flawed for one reason: *humans* were assessing the colors of birds. As you now appreciate that we see the world differently from most mammals, we also see the world differently than birds see it, or one another. Many birds possess ultraviolet vision, and see colors that we cannot, and this capability plays many roles—in mating, finding food, and even the feeding of young nestlings.

Many birds have tuned an opsin to detect ultraviolet light, and have evolved body markings that reflect light in the ultraviolet part of the spectrum (color plates D–K). Ultraviolet light has wavelengths less than 400 nm, which is shorter than violet, and is detected by the SWS opsin. Our SWS opsin is tuned to 417 nm. Various species of birds have SWS opsins that are tuned to wavelengths around 370 nm and can see in the ultraviolet. Other birds have SWS opsins tuned to the violet range of the visible spectrum, with absorption maxima around 405 nm, and those birds, like us, cannot see in the ultraviolet. Again, molecular studies in the laboratory have been able to pinpoint the precise changes in bird SWS opsins that determine whether the particular species is sensitive to violet or ultraviolet light.

One specific site, at position 90 in the bird SWS opsin, shows a perfect correlation with violet or ultraviolet vision, depending upon the amino acid found there. Birds with the amino acid serine in this position see in the violet range, birds with the amino acid cysteine in this position see in the ultraviolet range (figure 4.7). Furthermore, Shozo Yokoyama, one of the leading scientists studying color vision, and his colleagues at Emory University have shown directly that if serine is replaced with cysteine, a violet pigment becomes ultraviolet-sensitive, and if cysteine is replaced with serine in an ultraviolet pigment, it

Species	Order	Amino acid sequence at position 90	Opsin tpye
Osprey	Ciconiiformes	S	Violet
Herring gull	Ciconiiformes	C	Ultraviolet
Chicken	Galliformes	S	Violet
Hooded crow	Passeriformes	S	Violet
Common starling	Passeriformes	C	Ultraviolet
Woodpecker	Piciformes	S	Violet
Grey parrot	Psittaciformes	C	Ultraviolet
Ostrich	Struthoniformes	S	Violet
Common rhea	Struthoniformes	C	Ultraviolet

FIG. 4.7. **The evolution of ultraviolet vision in birds.** The detection of violet or ultraviolet wavelengths by bird SWS opsins depends critically on whether the amino acid at position 90 is serine (S) or cysteine (C). Changes at this position have occurred at least four times in different bird orders. *Figure by Jamie Carroll.*

becomes violet-sensitive. Just this single amino acid difference shifts the absorption maximum by 35 to 38 nm, a very dramatic shift. These studies show that a single change may alter the function of the SWS opsin, and therefore the evolution of either a violet- or ultraviolet-detecting opsin could be a relatively simple, one-step process.

The different birds that possess ultraviolet vision belong to nine different families in four different orders. Based upon their relationships to one another, it appears that ultraviolet vision has evolved at least four separate times in birds. All of the bird orders that have ultraviolet-sensitive species also contain violet-sensitive species, so it appears that the mutation that changed the serine to cysteine in a violet pigment happened multiple times. Here again, in ultraviolet vision, evolution has repeated itself. This is compelling evidence that selection has acted on opsin genes. In this case, it may well be that the nature of the selection was sexual, as there is abundant evidence now that mating preference in ultraviolet-sensitive species is affected by colors and patterns that are visible only in the ultraviolet range (color plates D–K).

For example, in starlings, female mating preference is determined by ultraviolet plumage colors, not by the plumage in the color range visible to humans. This was discovered by placing test birds in an apparatus in which the wavelengths of available light could be filtered. Females ranked males differently when UV light was available, and showed a preference for males that had ultraviolet markings on their throat feather tips.

Similarly, the male blue tit differs from the female in the ultraviolet reflectance of the blue feathers on the crest of their heads (plates D–G). In laboratory experiments, females prefer males with the brightest ultraviolet-reflecting crests. This prompted the University of Bristol researchers who did the study to declare, in the title of their paper, "Blue tits are really ultraviolet tits."

The utility of ultraviolet vision goes beyond mating preferences. Recently, it was shown that the mouths of chicks from eight different birds species are highly reflective in the ultraviolet band, particularly along the rim of their mouths. This means that parents returning to the dark nest can see the mouths of their nestlings. Furthermore, it appears that the stronger nestlings reflect more UV light, so in the competition among nestlings, the fittest appear to stand out by virtue of these UV-reflecting mouths, which are presented to a feeding parent.

There is also evidence that ultraviolet vision is used in hunting prey. Blue tits use their UV vision to help detect caterpillars that are camouflaged with visible colors. Kestrels, a type of raptor, hunt voles by honing in on the voles' scent marks, which are UV-reflecting, to locate areas of high prey density.

Ultraviolet vision is by no means limited to birds. Some fish, amphibians, reptiles, and mammals, such as bats, are known to use ultraviolet vision, and each UV-sensitive species has an SWS opsin tuned to around 360–370 nm. The widespread evolution and use of this capability suggests that it can play many roles. This is a general theme in evolution, that one innovation creates the opportunity to evolve additional innovations. I will close this chapter by returning to the colobus monkey, and an additional set of innovations it evolved that built upon its talent for identifying the most nutritious leaves in the forest. It is another very clear case of making something new from an "old" gene.

Ruminating Monkeys

While most primate diets focus on fruits and insects, colobus monkeys have specialized as leaf-eaters. And similar to those eaten by familiar barnyard ruminants, the leaves are fermented by bacteria that live in the foregut, one of the chambers of the colobus's multichambered digestive system. Just like cows and other ruminants, colobus extract the nutrients from that stew by breaking down and digesting the bacteria with a variety of enzymes. One such important enzyme, ribonuclease, is made in the pancreas and secreted into the small intestine, and it breaks down RNA. The ribonuclease helps to harvest the large amounts of nitrogen in the RNA of the fermenting bacteria. The colobus and conventional ruminants have greater amounts of ribonuclease in their pancreas than other mammals, which raises the question of how this digestive enzyme has evolved.

The key discovery was that, while most mammals and other monkeys have only one copy of the pancreatic ribonuclease gene, the colobus monkey has three such genes. The ribonuclease gene was duplicated in the evolution of these monkeys. Detailed study of the genes and the enzymes they encode by Jianzhi Zhang (of the University of Michigan) has revealed that while one gene has continued to produce an enzyme virtually identical to that found in nonruminating monkeys, the two "new" genes have undergone several changes that have fine-tuned them to the digestive system and needs of the colobus.

Overall, there are ten and thirteen changes in the protein sequences of the two "new" enzymes. One marked difference in the behavior of the new enzymes is that their optimal activities are in a more acidic environment than the optimal activity of the "old" enzyme. This difference correlates with differences in the digestive process between colobus and other primates.

This is a good correlation, but that is not the only evidence that natural selection has acted on ribonuclease evolution. More compelling evidence comes from analysis of the DNA and protein sequence of the new ribonucleases. Recall from the last chapter that most changes in DNA are synonymous. There is a much higher ratio of nonsynonymous to synonymous changes (about 4:1, whereas other proteins show a ratio of about 1:5) in the new ribonuclease genes. This is good evidence of natural selection acting to favor these specific changes in the protein.

Inventing and Adapting

The evolution of color vision and ruminant enzymes are but just a few of a very large number of examples of how genetic information is expanded and fine-tuned as species adapt to ecological niches. The duplication of genes and their fine-tuning by selection are pervasive in nature. Most of our genes belong to families that have expanded in the course of evolution. The accidental duplication of genes, or groups of genes, occurs quite frequently. In fact, there is considerable variation among individual humans in the number of copies of a variety of genes.

Because duplicate genes are initially redundant, only a fraction of duplicate genes will be preserved and undergo the sorts of functional changes I have described in opsins and ribonuclease. The retention and fine-tuning of genes is a species-specific process and depends on chance, selection, and time. Differences in the fate of genes contribute to the differences in gene number between species and, more important, physiological and other differences among species. We have seen that shifts in animal lifestyles, such as living in the ocean or specializing on a leaf diet, are accompanied by telltale changes in the genes involved in these shifts.

It turns out that the gain and fine-tuning of genes is one face of evolutionary adaptation. There is another side to the story, also written in DNA. When species' lifestyles evolve away from those of their ancestors, some genes' functions become dispensable and they start to decay. When we catch them in the act of decaying, it can tell us a lot about how a species has changed, and that will be the focus of the next chapter.

Coelacanth. *Photograph from JAGO submersible, Jurgen Schauer and Hans Fricke.*

Chapter 5

Fossil Genes:
Broken Pieces of
Yesterday's Life

. . . .

Nature is, after all, the only book that offers important
content on every page.

—Johann Wolfgang von Goethe

IT WAS A MAGNIFICENT EARLY CHRISTMAS PRESENT.
Midmorning on December 22, 1938, Marjorie Courtenay-
Latimer received a message from the manager of the local
fleet—the *Nerine* had docked and it might have some fish
for her collection. Miss Latimer was the first curator of the
East London Natural History Museum, in the Cape
Province of South Africa, and she was busy that day trying
to put together a dinosaur skeleton she had excavated, not
to mention getting ready for the holiday.

The fleet manager did not call very often, so she
decided to set aside her work and go down to the dock.
Hitching up her cotton dress, she boarded the trawler and
surveyed the stinking pile of sharks, sponges, and other
familiar creatures lying out in the heat of the sun. She was
about to return to the museum when it caught her eye. As

she pulled away a bunch of carcasses she saw "the most beautiful fish I had ever seen. . . . It was five feet long and a pale mauve-blue with iridescent markings."

It was also unlike any other fish she had ever seen. It was covered in hard scales, had four limblike fins, and a strange puppy-dog tail. She knew it had to be preserved. The fish weighed 127 pounds, so getting the dead and decomposing creature back to the museum was no small task. It took quite a bit of persuasion for a taxi driver to allow it into his trunk.

Once back at her post, she showed off her prize to the museum's director. He promptly dismissed it as a rock cod. Miss Latimer, virtually self-taught in natural history, thought differently, but none of her reference books helped her to identify the rotting hulk on her examination table. Miss Latimer decided to seek outside assistance in the form of Dr. J. L. B. Smith, a chemistry lecturer and amateur ichthyologist at Rhodes University, one hundred miles away. When she could not reach Smith by phone, she sent him a letter the very next day, enclosing a description and a drawing of the fish.

Smith did not get the letter until after the new year. He was recovering from an illness, and when he finally had the chance to study it he was bewildered. "And then a bomb seemed to burst in my brain, and beyond that sketch and the paper of the letter I was looking at a series of fishy creatures that flashed up as on a screen, fishes no longer there, fishes that had lived in dim past ages gone, and of which only fragmentary remains in rocks are known."

Smith immediately wired Miss Latimer: MOST IMPORTANT PRESERVE SKELETON AND GILLS FISH DESCRIBED.

Miss Latimer had managed in the meantime to get a taxidermist to preserve what he could.

Smith was positively riled by a possibility that his brain kept telling him was impossible. Yet the sketch, and then some scales he received later, told him that this fish was a coelacanth, a member of a group of fish with paired fins thought to be closely related to the first four-legged vertebrates *and believed to be extinct since the end of the Cretaceous period 65 million years ago.*

Smith finally got the chance to see the fish in person, which removed all doubt—as well as his faculties! He wrote, "I forgot everything else and just looked and looked, and then almost fearfully went close up and touched and stroked [it]."

Smith named the fish *Latimeria chalumnae*, in honor of Miss Latimer, and the river near where it was caught. It would be fourteen years before Smith or anyone else saw another coelacanth (and on that second occasion, Smith wept). Many more coelacanths have since been found in recent decades, including a second species discovered off Indonesia.

The coelacanth holds a special place in natural history. It is the only living member of an ancient tribe, with body features that link it to distant ancestors that lived 360 million years ago. The coelacanth has thus been dubbed a "living fossil."

In this chapter, we are going to uncover a different kind of fossil, one found in living species, that provides links to distant ancestors and former ways of life. These are *fossil genes.*

We have seen that shifts in species lifestyles, from land to water, from seeing visible hues to ultraviolet colors, and from eating fruits and insects to ruminating on leaves, involve the formation and fine-tuning of new genes. Here, we will see that such shifts also leave their traces in the form of genes whose use and function have been abandoned. Fossil genes reside in DNA much in the same way that fossils reside in sedimentary rock, and the text of the genes similarly breaks apart and erodes away over time. For information in DNA, the cardinal rule illustrated by fossil genes is "use it or lose it." The decaying text of fossil genes is evidence of the relaxation of natural selection and is very specific to individual genes and to particular species. We'll see how these broken pieces of yesterday's code reflect the adaptation of species, including humans, to new ways of life. I'll begin with coelacanth DNA and fossils of some now very familiar genes, and then work my way toward some examples of gene fossilization on a massive scale.

Shifting Habitats and Fossil Opsin Genes

Great fascination with the coelacanth has inspired expeditions to observe the animal in its native habitat. It has been seen from submersibles and by divers in deep underwater caves off the Comoros Islands and in waters around South Africa. The coelacanth retreats to these caves during the day and cruises slowly over the ocean floor at night to feed. At a depth of 100 meters or more, only dim blue light reaches the coelacanth in its native environment.

The coelacanth's lifestyle and its unique status have prompted interest in its visual system and opsin genes. Curiously, the coelacanth has a dim-light rhodopsin but no MWS/LWS opsin genes, the type other fish and we humans use for red-green vision. Because fish, mammals, and most other vertebrates have at least one version of this opsin, we know that coelacanth ancestors also had this gene, so somewhere along the line of coelacanth evolution the MWS/LWS opsin gene was lost. The loss of this gene raises a very general question: How and why is a gene that is so useful to some species lost in others? We can get a very good picture of the process of gene loss from another opsin gene that, while still in the coelacanth's DNA, is slowly eroding away.

The coelacanth has one SWS opsin gene. Remember that this is the short wavelength opsin used in humans and birds to detect the color violet, and in various species to see in the ultraviolet range. However, the code of the coelacanth SWS opsin gene contains many changes that disrupt its text. For instance, at positions 200–202 in the DNA code, where the mouse and other species have the three DNA bases CGA, the coelacanth has TGA. This change from a C to a T may seem like a tiny difference, but in this case it is a whopper. The letters TGA are a "stop" triplet that functions as a period to terminate the translation of the remainder of the SWS opsin text. This abolishes the coelacanth's ability to make a functional SWS opsin protein. Elsewhere in the code of the gene there are other deletions and changes that severely disrupt the opsin text. The many disruptions in the coela-

Dolphin	TTT	*TT	CTG	TTC	AAG	AAC	AT*	***	TTG
Cow	TTT	CTT	CTG	TTC	AAG	AAC	ATC	TCC	TTG

FIG. 5.1. **A fossil opsin gene in a dolphin.** Short portions of the sequences of a dolphin and a cow SWS opsin gene. The shaded regions show the positions of deleted bases (asterisks) that disrupt the code of the dolphin gene. *Figure by Jamie Carroll.*

canth opsin code tell us that it has become nonfunctional—it is a fossil gene (biologists call these "pseudogenes," but I will stick with the "fossil" label). It was functional in ancestors of the coelacanth, but no longer works in the fish today. The gene is still recognizable by the fragments of its parts. But, because it is not functional, it will continue to accumulate additional mutations and deletions that will erode it further, and will eventually erase it from the DNA forever, just as the coelacanth's MWS/LWS opsin was erased (and one of the globin genes was erased in the icefish ancestor I described in chapter 1).

You are probably wondering by now why a good gene would be allowed to decay. Are fossil genes a rare kind of mistake just found in weird animals like coelacanths and icefish? Before I explain further, I will mention one more illuminating example.

Inspection of the SWS opsin gene of dolphins and whales reveals that, just as in the coelacanth, these cetaceans' SWS opsin gene has become a fossil. For example, the SWS gene of the bottlenose dolphin is missing one base at one position and four bases at another position near the beginning of the text (figure 5.1). The missing bases throw off the three-base-at-a-time decoding of the gene's text, shifting the reading frame and making the gene nonfunctional. Examination of other cetaceans reveals a number of changes in their SWS opsin gene that have rendered them nonfunctional. All dolphins and whales have a fossil SWS opsin gene.

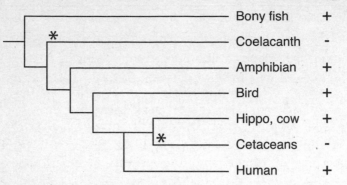

Functional
SWS opsin?

Bony fish   +

Coelacanth   -

Amphibian   +

Bird   +

Hippo, cow   +

Cetaceans   -

Human   +

✳ Gene fossilization

FIG. 5.2. **The same opsin gene has been fossilized twice.** The distribution of different mutations found in the coelacanth and cetacean SWS opsins and the evolutionary relationship of these species indicates that the SWS opsin was fossilized at least twice (asterisk). *Figure by Jamie Carroll.*

So, here, we have two examples of fossil SWS opsin genes, in the coelacanth and cetaceans. Is there anything in common or any connection between these animals that could explain why their SWS opsins are fossilized?

The first thing we can say is that the genes were fossilized independently. We know this from where these animals sit in the evolutionary tree of vertebrates (figure 5.2). The coelacanth belongs to a primitive group of fish that split off from a line that gave rise to four-legged vertebrates. Because amphibians, reptiles, birds, and many mammals have intact SWS opsin genes, we know that the fossilization of the coelacanth gene occurred during the evolution of the coelacanth lineage. Because hippos and cows, two close relatives of dolphins and whales, have functional SWS opsins, while all cetaceans do not, we can deduce that the function of the cetacean SWS opsin was lost in an ancestor of all dolphins and whales. The lack of functional SWS opsin in modern dolphins and whales is due to inheritance of this fossil gene from ancestors that lived more than 40 million years ago.

The best explanation for why the gene became fossilized comes from consideration of the animals' ecology. Surely, there must be some link to not using SWS opsin and these animals living in a marine habitat. Dolphins and whales are fully aquatic and belong to the only order of mammals that lacks the potential for any form of color vision (because its members have only one cone opsin, whereas most other mammals have two). As we saw in the last chapter, dolphins have tuned their dim-light-type rhodopsin to the bluer range of the light spectrum. The coelacanth is also a deep-dwelling animal. It, too, apparently has no use for color vision. The ecological rationale for the loss of the SWS opsin function is that it became dispensable to the ancestors of these species.

The dispensability of SWS opsin explains what we see happening in the DNA code. If the opsin is no longer needed, then natural selection, which would normally preserve the opsin's text, is *relaxed*. When natural selection is relaxed, there is no mechanism for purging genes of mutations that disrupt their function. The random process of mutation assures that *all* genes experience mutation. Most of the time, disruptive mutations are purged by the competitive process of natural selection because individuals and offspring bearing them are less fit. But when a trait is no longer under selection, in this case because of a shift in habitat, the genes that were essential in one lifestyle can become dispensable, and mutations can accumulate in them.

Use it or lose it.

In more formal terms, *fossil genes are exactly what we would predict to evolve as a consequence of the continuing action of mutation, over time, in the absence of natural selection.* The unmistakable signature of genes' dispensability is the accumulation of text-disrupting mutations. The resulting fossil genes are thus marks of changes in lifestyle from those of ancestors, and when we can spot and track fossil genes, these are valuable clues to reconstructing natural history.

I will highlight several more examples of opsin gene fossils linked to other kinds of changes in lifestyle and then expand the discussion of fossil genes to more genes, and many more species.

Living in Darkness

One leading theory to explain the reduction in the number of opsin genes and the loss of full color vision that occurred in the ancestors of mammals is that early mammals were nocturnal, and that color vision was dispensable for these rodentlike creatures' lifestyles. Nocturnality has evolved repeatedly in mammals, so one way of testing this theory, and the general idea of gene fossilization linked to shifts in lifestyle, is to examine more recent species with markedly different lifestyles.

The owl monkey, for example, is the only nocturnal species among higher primates (figure 5.3). And sure enough, examination of its SWS opsin gene reveals that it, too, has accumulated mutations that render it nonfunctional. Just sixty bases into the code for the opsin protein, the owl monkey has a mutation that changes a "TGG" into "TGA," another case of a stop triplet that terminates the translation of the remainder of the gene's text. All of the diurnal (daylight-active) relatives of the owl monkey have an intact SWS opsin gene, so this is pretty good evidence that the shift to nocturnality relaxed selection on the SWS opsin gene.

This idea about the link between nocturnality and SWS opsin can be further tested by examining the opsin genes of nocturnal prosimians. The prosimians are a primitive primate group that includes lemurs, tarsiers, bush babies, and the lorises. Lemurs include both nocturnal and diurnal species but the slow loris and bush baby are strictly nocturnal (see figure 5.3). Examination of their SWS opsin genes reveals, yet again, that the gene has become a fossil in these species. Each species gene has a big chunk of code missing near the beginning of the gene that obliterates the ability to make the opsin. Since the deletion is in exactly the same position and is exactly the

FIG. 5.3. **Nocturnal or subterranean mammals with fossilized opsin genes.** The owl monkey (top left; photograph by Greg and Mary Beth Dimijian), bush baby (bottom left; photograph by B. Smith, Cercopan, Nigeria), slow loris (top right; photograph by Larry P. Tackett, www.tackettproductions.com), and blind mole rat (bottom right; photograph by Tali Kimchi) all have fossilized SWS opsin genes, a result of their adaptation to nocturnal or subterranean lifestyles.

same size in each species, this indicates that the fossilization of SWS opsin first occurred in a common ancestor of the slow loris and bush baby, and was inherited by these species.

So far, so good. It appears that shifts in the light environments that species live in are well correlated with the loss, or retention, of color vision genes. Just one more test: How about when animals go underground?

The blind mole rat is a rodent that possesses the most degenerated eyes of any mammal (see figure 5.3). The fossil record suggests that this group of animals evolved from an aboveground ancestor that had normally proportioned eyes. The evolution of the mole rat's lifestyle has been accompanied by many changes in anatomy and physiology. Its eyes are so tiny that they cannot detect images. And even if the eyes themselves worked, seeing would still be difficult because they are located entirely under the skin and covered by a layer of fur. However, the blind mole rat can tell the time of day—its eyes have retinas that detect light and help maintain its circadian clock, which regulates its daily biorhythms.

Examination of the blind mole rat revealed two intact opsin genes, a red-shifted MWS/LWS pigment that is tuned to detect the light received through the subcutaneous eye and a dim-light rhodopsin. Clearly, despite its atrophied vision, selection remains on these genes, apparently in order to run the animal's biological clock. However, its SWS opsin gene is a fossil and contains numerous mutations that disrupt the text of the code for making the SWS opsin protein.

I have now described five separate cases of SWS opsin gene fossilization—in coelacanths, cetaceans, owl monkeys, the slow loris and bush baby, and the blind mole rat. In each case, the fossilization of the gene is correlated with the habitat in which the species live. In each case, the exact lesion in the SWS opsin gene is different. This fact, and the facts that the species belong to different parts of the evolutionary tree, and that close relatives of these animals have functional opsins, demonstrate that the fossilization of the SWS opsin has occurred independently and repeatedly by different mutations and at

different times in history. This is overwhelming support for the fundamental prediction that relaxed selection on a gene will lead to its
decay. Furthermore, in all these species, other kinds of opsins are
intact and functioning, which demonstrates that the decay of genes is
highly selective.

The frequent loss of SWS opsin and its correlation with shifts in
lifestyle is a very loud hint that the fossilization of genes is a frequent signature of evolutionary change. Let's now turn to the human
genome for some fossil signatures of how we are different from our
ancestors.

Can't You Smell That Smell?

We have now seen many examples of how shifts in habitat are associated with the adaptation and loss of visual system genes. Other senses
are also vital to animal behavior and survival, particularly the sense of
smell. One walk in the park with a dog provides many examples of
how their "view" of the world is shaped by their acute sense of smell.

Many other mammals also have powerful senses of smell, which
are used to find food, identify mates and offspring, and detect danger.
For a long time, it was a mystery how different odors could be
detected and discriminated. In 1991, Linda Buck and Richard Axel
discovered a family of genes that encoded odorant receptors. It
turned out that the so-called olfactory receptor genes are the largest
family of genes in mammal genomes. Mice have about 1400 of these
genes in a genome of about 25,000 genes. It was discovered that the
specificity of olfaction is due to each sensory neuron in the olfactory
system producing just one, or in some cases a few, of these many
olfactory receptors, each capable of detecting different groups of
odorants. How a given chemical "scent" is perceived depends on the
combination of receptors that detect it. For cracking the mystery of
the genetics of smell, Buck and Axel were awarded the 2004 Nobel
Prize in Physiology or Medicine.

The human olfactory genes have been studied in great detail, and it turns out that compared to the mouse, our olfactory genes are nothing to brag about. About half of all our olfactory receptor genes are fossilized and incapable of making functional receptors. The contrast between humans and other mammals is most striking for one class of receptors encoded by the *V1r* genes. The mouse has about 160 functional V1r receptors, while we have only 5 functional *V1r* genes out of more than 200 in our genome. Our repertoire of olfactory receptor genes has gone to pot.

The extraordinary proportion of fossilized olfactory receptor genes suggests that we no longer rely on our sense of smell to the degree that our ancestors once did. Two questions immediately come to mind. First, why have we abandoned the use of such a large fraction of odor receptors? And second, when in evolution did this happen?

Clues to the answers to both questions emerge from studying the fraction of fossilized odor receptors in other primates and mammals. Yoav Gilad and colleagues at the Weizmann Institute of Science in Rehovot, Israel, and the Max Planck Institute for Evolutionary Anthropology in Leipzig, Germany, surveyed the olfactory gene repertoire in apes, Old World monkeys, New World monkeys, and lemurs and compared it to that of the mouse. They found a striking correlation between the proportion of fossilized olfactory receptor genes and the evolution of full color vision. In mice, lemurs, and New World monkeys that lack full color vision about 18 percent of olfactory receptor genes are fossilized. But in the colobus and Old World monkeys about 29 percent of receptor genes are fossilized, and in nonhuman apes such as the orangutan, chimp, and gorilla, this rises to 33 percent. Finally, in humans they found that 50 percent of this repertoire was fossilized. The fraction of fossil olfactory receptor genes is significantly higher in all species with full color vision. This suggests that the evolution of trichromatic vision—which allows these primates to detect food, mates, and danger with visual cues—has reduced their reliance on the sense of smell. Relaxed selection on the olfactory receptor genes in trichromatic species has allowed the genes' codes to decay. Conversely, in animals that rely heavily on their sense of smell, the fraction of intact genes is much higher.

There are other physical, behavioral, and genetic signs of this reduced reliance on the sense of smell in humans and other primates. The vomeronasal organ, a cigar-shaped sensory organ located toward the front of the nasal cavity, detects pheromones in most land vertebrates. But this again is greatly reduced in humans and higher primates in comparison with other species. The V1r receptors I mentioned above play a critical role in pheromone detection. So it appears that we are also less reliant on pheromones than other mammals, again perhaps because our ancestors relied more on visual signals in mating and other behaviors.

Because the vomeronasal organ and V1r receptors are so reduced in humans and other higher primates, we might also predict that other machinery involved in the transmission of information from the olfactory system would degenerate. This is exactly the case. Another gene that plays a specific role in vomeronasal organ function is called *TRPC2* and encodes a protein that regulates the trafficking of ions in sensory cells. In mice, TRPC2 is fully functional and required for normal behaviors in response to pheromones. But in humans, and all of the higher primates with trichromatic vision and larger numbers of fossilized olfactory receptors, the *TRPC2* gene contains a bunch of mutations that have rendered it a fossil gene.

The fact that different kinds of genes with different jobs in the nose are fossilized is a very striking and satisfying fulfillment of the prediction about the effect of relaxed selection on species traits. That is, when an entire organ or process falls into disuse, different genes responsible for different steps in a process may all experience relaxed selection and then undergo fossilization. The evolution of the vomeronasal organ and parts of its machinery raises the possibility that whole pathways of genes may become dispensable, decay, and eventually disappear. This is exactly what we see happening in some species, sometimes on a massive scale. I'll describe two examples from other kingdoms that vividly illustrate how evolution throws away what is no longer useful.

Use It or Lose It

Yeasts and other fungi play important roles in human affairs. We use yeasts to ferment beer and wine and in bread-making, while fungi were the source of the first antibiotics. Because it is so easy to culture, the baker's and brewer's yeast *Saccharomyces cerevisiae* has been a favorite laboratory organism for many years. Through experiments with yeast, much has been learned about how cells grow and divide, how genes are used, and the biochemistry of life.

But there are many more yeast species than just good old baker's yeast. Under the microscope most of them look very similar. However, there are often detectable differences in species' abilities to metabolize nutrients and to grow in different environments. The breakdown of many nutrients into their useful components often occurs through a sequence of steps, or a *pathway*. One of the best studied nutrient pathways in any living organism is the galactose pathway of baker's yeast. Most organisms use the sugar glucose as an energy source. When it is not available, either stored sugars (starches) or alternatives must be used. Baker's yeast can utilize the sugar galactose as an alternative source because it can convert galactose into a usable form of glucose through a series of enzymatic steps. These steps require four different enzymes, encoded by four different genes. Furthermore, to ensure that the yeast makes these enzymes only when they are needed and galactose is available, three other proteins control the making of the enzymes. Altogether then, seven genes are devoted to running the galactose pathway in baker's yeast.

Most close relatives of baker's yeast can also utilize galactose, except for one. This species, *Saccharomyces kudriavzevii* (try saying that quickly several times), was discovered on decaying leaves in

Japan, unlike the sugar-rich places where most other yeast species live in the field. When my graduate student Chris Hittinger looked at *S. kudriavzevii*'s seven genes of the galactose pathway, he quickly found out why this species can't utilize galactose—each gene was shot to hell. Each of the seven genes had various-sized chunks of code missing that obliterated the integrity of their text.

There is a very striking contrast between the state of the seven galactose genes and their immediate neighbors in the *S. kudriavzevii* DNA. The neighboring genes are perfectly intact, just as they are in baker's yeast and other related species. If one thinks of each gene's code as about one large paragraph of text, the text of each galactose gene's code in *S. kudriavzevii* is obliterated in many places, but the preceding and following paragraphs encoding other genes are untouched. This pattern reveals how exquisitely specific the fossilization of genes is. A gene that is no longer needed or used accumulates many mutations, while the codes of neighboring genes that are used are perfectly maintained. The fate of genes in *S. kudriavzevii* demonstrates how natural selection maintains what is needed, but it cannot maintain what is no longer needed. This species adapted to living on other sugar sources, and its galactose pathway was no longer needed and fell into disuse. Without the constant surveillance of natural selection to purge the galactose genes of inactivating mutations, the genes fossilized and are well on their way to being erased.

The selective decay of seven functionally related genes is a great example of how a substantial number of genes are abandoned by the relaxation of natural selection, but it pales in comparison to what has happened in some other microbes, such as the species *Mycobacterium leprae*, the pathogen responsible for the disease leprosy.

Sequencing of the *M. leprae* genome revealed that it contains about 1600 functional genes, and almost 1100 fossil genes—an enormous fraction of dead genes, and far greater than that of any other known species. *M. leprae* is closely related to *M. tuberculosis*, the species responsible for pulmonary tuberculosis. But *M. tuberculosis* has about 4000 intact, functional genes and only about 6 fossil genes. The difference between the two species reveals that *M. leprae* has fossilized or lost about 2000 genes in the course of its evolution. What explains the vast difference in the numbers of functional and fossil genes between the two organisms?

M. leprae has a very different lifestyle from its cousin. It can live only within cells of its host. It resides within cells called macrophages and infects cells of the peripheral nervous system, whose eventual destruction leads to the physical disfigurement typical of the disease. It is the slowest-growing bacteria of all known species (it takes about two weeks to divide, while the *E. coli* in our gut can divide every twenty minutes). Despite decades of effort, it has never been grown on its own in the laboratory. The specialization of living within host cells has allowed *M. leprae* to rely on the host for many metabolic processes. With the host cell genes doing much of the work, this has relaxed selection for the maintenance of many *M. leprae* genes. The massive decay of genes seen in *M. leprae* has also occurred in other intracellular parasites and pathogens. Species with vastly reduced numbers of functional genes demonstrate that a large fraction of all genes can become dispensable upon shifts in organisms' lifestyles.

The fossilization of individual genes, sets of genes in pathways, or larger groups of genes in species has important consequences for the future evolution of their descendants. Because decaying genes generally accumulate multiple defects, their inactivation cannot be easily reversed. This means that the loss of gene functions is generally a one-way street. Once gone, these functions will not return. Just as new species of icefish will not have or use hemoglobin, species that evolve from *S. kudriavzevii* will not be able to use galactose. Gene fossilization and loss imposes constraints on the future direction of evolution in lineages.

"Use it or lose it" is an absolute rule imposed by the fact that surveillance by natural selection acts only in the present—it cannot plan for the future. The downside to this rule is that if circumstances change, even over long periods of time, species that have lost particular genes will not have those genes available to adapt to new circumstances. This may be an important factor in the success or extinction of species. Keep in mind that biologists think that over 99 percent of all species that ever existed are now extinct.

Cause or Effect?

The prevalence of fossil genes offers powerful new means of viewing the process of evolution. However, it also raises questions about cause and effect. Is the fossilization of genes a cause of evolutionary change brought about by natural selection, or is fossilization largely an effect—a by-product of natural selection for other features? The answer appears to be that it can be either, depending upon the circumstances. I will explore this issue with examples of recently evolved fossil genes in a flowering plant and in humans that are most likely examples of cause and effect, respectively.

In plants, flower colors are often used to attract pollinators, particularly bees or birds. There are many well-documented cases of evolutionary shifts in pollinator species. It is easy to imagine how changes in climate or the abundance of pollinators could select for variations in flower colors. Furthermore, because flowers that provide nectar to hummingbirds or bees may be visited by unwelcome pests, there is selection on other structural features of flowers that fine-tune flower anatomy to different pollinators. For example, bird-pollinated species tend to produce larger amounts of nectar and have narrow floral tubes while bee-pollinated species produce small amounts of nectar and have broad floral tubes.

In the morning glory genus *Ipomoea*, the ancestral flower color was blue or purple. This group is typically pollinated by bees, but one species, *Ipomoea quamoclit*, has red flowers and is pollinated by hummingbirds. The red color appears to be an adaptation for attracting hummingbirds.

The production of red, blue, or purple flower color is determined by an enzymatic pathway in morning glories. Beginning with a common precursor, different sets of enzymes produce either the blue and purple or the red pigments. Recently, Rebecca Zufall and Mark Rausher at Duke University showed that the pathway for making blue and purple pigments has degenerated in the red *I. quamoclit*. One enzyme in the blue/purple part of the pathway appears to be completely impaired while a second is altered such that it can contribute to red but not blue/purple pigment synthesis.

Because the evolution of red flower color is adaptive, and its evolution in the morning glory is most likely directly due to changes in these two enzymes, it appears that gene inactivation in this instance is a cause of evolution. Natural selection may well have favored the inactivation of the blue/purple-promoting enzymes and the evolution of red coloration, as opposed to the enzymes' disuse evolving as a byproduct of selection for some other trait.

More often, however, gene inactivation and fossilization is likely to be a consequence of relaxed selection on genes, where gene inactivation is among the last in a series of changes that have rendered a gene dispensable. This is probably the case for the opsin genes I have described and for a fascinating case of gene fossilization that occurred specifically in the human lineage, after our line split off from our last common ancestor shared with chimpanzees.

THE HUMAN GENE that was fossilized is called *MYH16*. In humans there is a two-base deletion in the *MYH16* code that disrupts the proper reading of its text (a relevant portion of the text is shown in figure 5.4; the deletion is indicated by asterisks). In chimps, gorillas, orangutans, and macaques the gene is perfectly intact.

In other primates, the MYH16 protein is made in a subset of muscles, particularly the very prominent temporalis muscle that extends over most of the area of each side of the temporal region of the skull. The large temporalis muscle is involved in movements of the large jaws in apes involved in chewing. In humans, the temporalis region

Human	ATG	ACC	ACC	CTC	CAT	AGC	**C	CGC
Chimp	ATG	ACC	ACC	CTC	CAT	AGC'	ACC	CGC
Gorilla	ATG	ACC	ACC	CTC	CAT	AGC	ACC	CGC
Macaque	ATG	ACC	ACC	CTC	CAT	AGC	ACC	CGC

FIG. 5.4. **Fossilization of a human muscle gene.** A short portion of the sequence of the *MYH16* myosin muscle gene is shown. In humans, a deletion of two bases disrupts the code of the gene (asterisks) and is associated with the reduction of two muscles involved in chewing, which are massive in our ape relatives. *Figure by Jamie Carroll, based on data of H. H. Stedman et al. (2004),* Nature *428:416.*

and muscle are much reduced in comparison with gorillas, chimps, and macaques. The MYH16 protein is a myosin that forms part of the large fibers within muscles that generate their force. Human temporalis muscle fibers are much smaller than those in our relatives, and smaller fibers make for smaller muscles.

The intriguing correlation between a mutation in a protein affecting muscle and fiber size and evolutionary changes in the temporalis muscle raises the question of whether the mutation was a cause of the muscle's reduction or a by-product of the muscle's reduction that occurred by other means. This is difficult to say for certain, but we can consider some additional evidence in weighing the alternative explanations. It is known that mutations in this class of proteins can have severe effects on muscles. If a primate with a large jaw (such as that of apes and our ancestors) lost this muscle in one step, it would not be able to chew. In order for the *MYH16* mutation to have played a role in the evolution of the temporalis muscle, we should think of scenarios in which the muscle mass was not lost all at once. I think that the explanation for the fossilization of *MYH16* is more likely to be similar to the history of globins in the icefish—that is, the muscle was becoming reduced by other genetic pathways and the fossilization of the *MYH16* was most likely a later event after the gene became dispensable.

Fossil Genes as Evidence Against "Progress" and "Design"

The examples described in this chapter demonstrate how the making of the fittest—whether speaking of ancient tribes of fish, magnificent dolphins, colorful flowers, slender-jawed humans, simple yeasts, or blind subterranean rodents—is not necessarily a "progressive," additive process. Modern species are not better equipped than their ancestors, they are mostly just different. They have often gained some coding information in their DNA and, as I have shown throughout this chapter, they have often lost some, or even many, genes and capabilities along the way.

The fossilization and loss of genes are powerful arguments against notions of "design" or intent in the making of species. In the evolution of the leprosy bacterium, for example, we don't see evidence that this pathogen was designed. Rather, we see that the organism is a stripped-down version of a mycobacterium, which still carries around over a thousand useless, broken genes that are vestiges of its ancestry. Similarly, we carry around the genetic vestiges of an olfactory system that was once much more acute than what we have today.

The patterns of gain and loss seen in species' DNA are exactly what we should expect if natural selection acts only in the present, and not as an engineer or designer would. Natural selection cannot preserve what is not being used, and it cannot plan for the future. The fossilization and loss of genes are exactly what is predicted to evolve in the absence of natural selection. Over time, chance mutations will accumulate, eventually disrupting the text of unused or unnecessary genes.

Furthermore, the repetition of gene fossilization in different ancestors of entirely different groups of animals is striking evidence that, when selection is relaxed on a particular trait, the same events will repeat themselves in DNA. The repetition of the independent fossilization of the SWS opsin gene described in this chapter is a profound demonstration of this principle. It is also a foretaste of the broader message of the next chapter, about the predictability and reproducibility of evolution in general, and the many amazing instances in which evolution has repeated itself.

Howler monkey, Costa Rica. *Photograph by Steven Holt.*

Chapter 6

Déjà Vu: How and Why Evolution Repeats Itself

· · · · · · · · · · · · ·

It is no great wonder if in the long process of time, while
fortune takes her course hither and thither, numerous
coincidences should spontaneously occur. If the number
and variety of subjects to be wrought upon be infinite, it
is all the more easy for fortune, with such an abundance
of material, to effect this similarity of results.

—Plutarch, *Life of Sertorius*

THE BUSH PLANE FLEW LOW ALONG THE RUGGED,
lush green Costa Rican coastline before veering inland to
make a perfect landing on a tiny strip near the Rio Sierpe.
We piled into a van and bounced our way through palm
plantations to the river landing, then hopped aboard a boat
and headed downstream. Flanked by thick mangrove jungle
as far as the eye could see, the wide river eventually poured
into the open Pacific Ocean. Pounding through rough surf
at the river's mouth, we landed on a beach at the edge of
Corcovado National Park on the Osa peninsula, one of the

last remaining large tracts of wilderness in Central America. After the long, spectacular journey we were completely exhausted and settled down for the night in the quiet of the rain forest.

Our well-earned slumber was shattered by the "dawn chorus"—the loud, low throaty calls of a raucous troop of howler monkeys cavorting overhead in the canopy. So much for the tranquillity of Nature.

Audible for up to three miles, the howlers have enlarged throats and extralarge voice boxes that enable them to broadcast their location to nearby, and not so nearby, troops (and tourists). The howler's call is unique, but the monkey has another attribute that sets it apart from the capuchin and squirrel monkeys, and all other New World monkeys with whom it shares the rain forest. The howler monkey has full trichromatic color vision.

We know from howler DNA that it acquired this sense in a similar way as its more distant Old World relatives, but at a different time in an entirely independent series of events. And what's more, the howlers feed on various sorts of tender young leaves, just as other trichromatic primates do. The independent repetition of monkey evolution does not end with these traits. The howler, like its distant African and Asian cousins, has a higher proportion of fossilized olfactory receptor genes than all of it closest relatives in the New World. It, too, has traded some of its sense of smell for seeing in color.

Astonishing.

The evolution of color vision, leaf-feeding habits, and loss of olfactory genes in howler monkeys occurred on a different continent, 20 to 25 million years after these traits evolved in the ancestor of African and Asian monkeys and apes. The natural history of these primates suggests that similar conditions in different parts of the world can favor the gain and loss of similar traits at different times in different species.

The howler monkey is a poster species for a pervasive phenomenon in nature called convergent evolution. In all sorts of animals, we see similar traits that evolved as independent inventions. The flippers of penguins, seals, and dolphins, for example, all serve a similar purpose in swimming, but each group of animals evolved from different ancestors that did not have flippers. The wings of pterosaurs, birds, and bats also evolved convergently; so did the similar body forms of ichthyosaurs and dolphins, and of snakes and legless lizards, to name just a few of many, many examples. The widespread occurrence of convergence is compelling evidence that, given a set of similar conditions, species often find similar "solutions" in adapting to these conditions.

In detail, however, many convergent structures differ in how they evolved. Bird, bat, and pterosaur wings differ in architecture, with different parts of the forelimb contributing the major portion of the wing surface in each group of animals. At the DNA level, such different structural details are expected to involve evolutionary changes in different genes. What makes the howler case so remarkable, as well as many of the other examples I will describe in this chapter, is that the recurring events in different species involve the same genes, and sometimes the very same letters of DNA code. We have already seen an example of such precise repetition in chapter 4, where I showed how exactly the same change had occurred at least four times in the short wavelength opsin of birds in the evolution of ultraviolet vision. And in chapter 5, I described five groups of animals in which this same opsin was inactivated by mutations and fossilized. The power of these examples is that they are documentary proof of the repetition of evolution at its deepest, most fundamental level.

They are just the tip of the iceberg.

In this chapter I will draw on many more wonderful examples of how evolution repeats itself. Several of the recurring adaptations I will describe are already familiar from previous chapters—the evolution of a new pancreatic enzyme in ruminants, the evolution of antifreeze in cold-water fish, and the loss of galactose pathway genes in yeast. I will also introduce some new traits, such as black or white body color, that have evolved repeatedly by similar means in vastly different species. These demonstrations of similar means of evolution in unrelated species provide overwhelming evidence of how natural selection, acting through variation in DNA, shapes the evolution of species.

The reproducibility of natural history raises the intriguing questions of how and why evolution repeats itself. I will explain later in the chapter that the answers lie in the interplay of chance, time, and selection and the arithmetic of some large numbers that determine the frequency of events in DNA and in Nature. Recurring events in Nature and the numbers that explain them provide the big "ahas" of this chapter and crystallize many of the key ideas of this book.

In order to understand how evolutionary convergence is revealed, let's begin with a more detailed look at the howler monkey and how we know for certain that the howler's traits evolved separately from those of Old World primates.

The Second Coming of Trichromatic Monkeys

The full color vision of howler monkeys was a surprising discovery made in a broad survey of New World monkeys. In principle, the presence of this capability in both howler monkey and Old World primates could be due to one of two alternative evolutionary scenarios. First, the howler monkey could have inherited full color vision from a common ancestor shared with the Old World primates. The second possibility, which I have already revealed is the correct one, is that howlers acquired full color vision in separate events from color vision evolution in Old World species.

How do we know the correct scenario?

When a trait is shared between any two species, we can distinguish whether it is likely to have been inherited from a common ancestor or to have evolved independently by considering the species relationships in an evolutionary tree. By mapping the presence or absence of a trait onto an evolutionary tree, the distribution of the trait reveals its evolutionary history. Each fork or branch point in the tree represents a common ancestor of each branch. If all species that share an immediate common ancestor both possess a trait, then it is most likely that the ancestor possessed that trait (figure 6.1A). If, however, the branches that join two

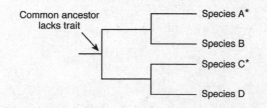

A. Inheritance from common ancestor

Common ancestor
with trait*

Species A*

Species B*

Species C*

Species D*

B. Independent evolution in two lineages

Common ancestor
lacks trait

Species A*

Species B

Species C*

Species D

FIG. 6.1. **Alternative evolutionary histories of shared traits.**
Species may share traits (asterisks) because they share a common
ancestor that also had the trait (A), or because they evolved the trait
independently after their separation from a common ancestor (B).
Drawing by Leanne Olds.

species include species that lack the trait, then the trait is most likely to
have been independently acquired in the two lineages (figure 6.1B).

Now look at the tree of Old World and New World primates
(deduced on the basis of the inheritance of SINES, LINES, and other
DNA sequences) in figure 6.2. The tree shows that the howler is most
closely related to other New World monkeys, all of which lack full
color vision. It is theoretically possible that an ancestor of New World
monkeys had full color vision and that all but the howler lost it.
However, this would require just one gain of color vision and very
many losses. The simpler explanation, the one that requires the fewest
evolutionary changes, is that the howler evolved from a dichromatic
ancestor and gained full color vision on its own.

Fortunately, we do not have to rely solely on interpretation of evolutionary trees to decide the truth of the matter. Gene duplications leave

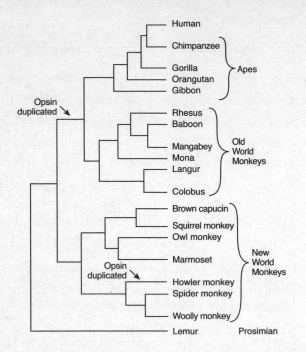

FIG. 6.2. **Full color vision evolved twice in primates.** Based on the distribution of color vision and the evolutionary relationships of primates depicted in this tree diagram, full color vision evolved twice (arrows) in primates—once in the common ancestor of apes and Old World monkeys and once in the howler monkey lineage. *Adapted from Gilad et al. (2004)*, PloS Biology *2 (1):e5.*

traces of their history in DNA. By looking at the actual arrangement and sequence of opsin genes in the DNA of Old World primates, the howler monkey, and other New World monkeys, telltale clues to each event emerge. In the DNA text, it is clear that the duplications of the opsin genes in Old World primates and the howler monkeys were different events. We know this because the size of the DNA region that was duplicated was different in each event. In Old World primates, the two duplicated genes share 236 base pairs of DNA text outside of the coding portion of each gene. The shared sequences indicate that 236 base pairs of adjacent sequence were duplicated in the making of the Old World

MWS opsin gene. However, in the howler monkey, the duplicated genes share a much larger region of text outside of each gene. This evidence is consistent only with the duplication of the opsin genes in howlers occurring as a separate event from their duplication in our ancestors.

One further piece of supporting evidence comes from the extent of sequence differences between the two duplicated genes. After a duplication event, because of the steady beat of mutation, each copy of a gene will accumulate changes. The older the duplication event, the greater the divergence will be between the texts of the two sister genes. In all Old World primates, the texts of the two opsin genes differ by more than 5 percent, while in the howler the two genes differ by just 2.7 percent. This indicates that the duplication of the howler genes occurred more recently than the duplication of the Old World genes. This conclusion is consistent with geological evidence of the more recent evolution of New World monkeys, after the separation of the South American and African continents.

The evolutionary convergence of howler monkey opsins extends beyond just the duplication of the genes. Recall that full color vision requires fine-tuning of opsins to different wavelengths and that three positions in the MWS and LWS pigments make the crucial difference between the wavelengths of light detected by the two opsins. The howler monkey MWS and LWS are tuned to the exact same wavelengths, and contain exactly the same amino acids, in these three key positions, as the MWS and LWS opsins of human and other Old World primate species. This means that three of the same changes have occurred in the evolution of the "new" MWS howler opsin and in the "new" MWS opsin in Old World primates.

All of the DNA evidence demonstrates that the steps taken in the evolution of howler vision and olfaction have followed the same path as that of Old World primates millions of years earlier. The duplication of the opsin gene, the fine-tuning of key sites in the opsins, and the fossilization of olfactory genes were repeated in the same order, and in several of the same details.

In the convergent evolution of color vision in primates and of ultraviolet vision in certain birds, we see somewhat closely related species acquiring similar characteristics, but convergence is not at all limited to such "cousins." Recall the story of the evolution of the dim blue rhodopsin of deep-sea eels and the beaked whale (chapter 4) that evolved the same three amino acids at key sites. Same story, different gene, and vastly different species.

The natural history of the opsins raises the general question: How often are similar traits in different species due to similar evolutionary events in DNA? Let's look at four cases where similar means, involving the same gene or genes, were used in the evolution of similar traits at different times in different species.

Similar Means to Similar Ends

Long before the colobus monkey evolved a foregut and the ability to ferment leaves, an ancestor of the more familiar barnyard ruminants such as cows, sheep, or goats evolved this capability. Are there any similarities between how rumination evolved in monkeys and cows? You bet.

Recall that one adaptation of the colobus monkey was a specialized pancreatic ribonuclease that breaks down nutrients in the fermenting stew of leaves and bacteria. The pancreatic enzyme evolved by duplication and fine-tuning of a gene that encoded a general ribonuclease. In cows, the same gene is also duplicated and fine-tuned to the environment of the cow gut. We know that the two events in the monkey and cow occurred separately because all ruminants have the duplicated ribonuclease gene, but ruminants' closest relatives, such as hippos and dolphins, and the colobus' closest relative have a single gene. The two types of ruminants could not have inherited the duplicated genes from a common ancestor.

Moreover, it turns out the African colobus monkey is not the only group of ruminating monkeys. In Asia, a separate group of monkeys evolved rumination. The spectacularly marked douc langur (color plate L), an endangered species that lives in Vietnam, Laos, Cambodia, and

China, also has duplicated ribonuclease genes. Jianzhi Zhang of the University of Michigan has found that the duplication of ribonuclease genes occurred at different times and produced different numbers of ribonucleases (three in Africa, two in Asia). However, in the ribonuclease enzymes, several of the exact same changes subsequently occurred. The probability of identical changes becoming established in both groups of monkeys by chance is minuscule. Rather, the parallel changes in the monkey enzymes in evolutionary history is a signature of natural selection acting to fine-tune the enzymes to the more acidic environment of the monkey's foregut.

The fossilization and loss of genes is also repeated. In the last chapter, I described how the yeast *S. kudriavzevii* has selectively lost the function of all seven genes dedicated to the metabolism of galactose. Three other yeast species, each belonging to a different genus and separated by millions of years of evolution, have also lost most or all of their galactose genes and cannot utilize the sugar. Based upon the evolutionary relationships among yeast, we can say for certain that there have been at least three (and probably more) times in evolution when these genes have been lost. In each case, it is likely that relaxed selection has allowed the genes to decay and to be erased.

The relaxation of natural selection also explains the repeated evolution of traits in cave-dwelling animals. There are many species of fish, for example, that live in caves and have lost their eyes and body color. Because these species belong to many different families that contain surface-dwelling, eye-bearing species, it is clear that the loss of eyes and pigmentation has occurred repeatedly. These cave fish present a perfect opportunity to find out whether similar outward appearances have a deeper, common cause.

Recently, Meredith Protas and Cliff Tabin of Harvard Medical School, Bill Jeffery of the University of Maryland, and several collaborators studied the evolution of the albino form of the Mexican blind cave fish (*Astyanax mexicanus*) (figure 6.3). This fish belongs to the same order as the piranha and the colorful neon tetra, but there are about thirty cave populations in Mexico that have lost the body color

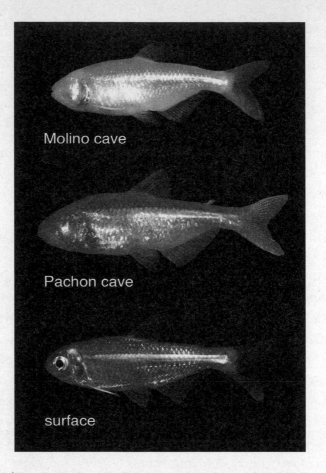

FIG. 6.3. **The evolution of albinism in blind cave fish.** While surface forms of the fish *Astyanax mexicanus* appear normal, cave populations of the fish have repeatedly evolved blindness and albinism, such as the two forms shown here from Molino and Pachon caves, through mutations in the same gene. *Photographs courtesy of Meredith Protas and Cliff Tabin, Harvard Medical School.*

of their surface-dwelling cousins. The researchers discovered that in the two populations they studied the same pigmentation gene was inactivated by a deletion of DNA text, but that the exact deletion was different in each population. This is definitive evidence that the different cave populations lost their color independently.

Albinism in cave fish is most easily explained by the relaxation of natural selection on body color; in the dark who cares what you look like? But in many other animals body color is important in mate choice, predator evasion, and other behaviors affected by sexual or natural selection. One of the most common body colors is black. Many species have some black coloration of their fur, scales, or feathers. Variation within species is also common, where either the different sexes or different populations vary in the amount of black coloration. In many cases, natural selection or sexual selection has operated on the same gene involved in vertebrate body color.

For example, lesser snow geese may be either white or "blue"; the latter appears so because of black pigment in their feathers (color plates M and N). The distribution of geese colors varies across their range, with blue individuals most common in eastern Canada and white individuals most common in the western part of their range, in eastern Siberia. The differences in color are important in mate choice. Young snow geese learn their parent's color at an early age and later prefer to mate with individuals of the same color. A single genetic difference is responsible for the two types. The gene responsible is called the melanocortin-1 receptor, or *MC1R* for short. The white snow geese *MC1R* gene differs from that of the blue snow geese at just the one triplet that encodes amino acid 85.

In other birds, variation in the sequence of the *MC1R* gene is also perfectly correlated with coloration. The dark form of the bananaquit has a single change in its *MC1R* gene that differentiates it from yellow birds. This single change is in a different position in the gene than the position that determines the white/blue color of snow geese. It is also different from the position that determines the light versus dark color of a third bird species, the arctic skua (plate O). In arctic skuas, too, plumage color affects mate choice and is under sexual selection. The dramatic color differences of the fairy wren are also determined by *MC1R* (plates P and Q).

The role of *MC1R* in body color evolution extends well beyond birds. Variation in *MC1R* is responsible for the difference between the orange and black color phases of the jaguar, the white versus black forms of the black bear in western North America, light and dark forms of various lizards, as well as coat color variation in domestic dog, horse, and cat breeds.

One of the best studied examples of the *MC1R* gene's role in evolution in the wild is in the rock pocket mouse of the southwestern deserts of the United States. In chapter 2, I used the known genetics of *MC1R* in mice and the evolution of light and dark forms of the pocket mouse to illustrate the interplay of chance mutations and selection and the timescale of evolution. These mice inhabit sandy desert and black lava outcrops in Arizona and New Mexico. Their alternate coat colors afford protection against predators when color-matched on different backgrounds. Michael Nachman, Hopi Hoekstra, and collaborators at the University of Arizona have shown that, in the Pinacate region, the dark form of the mouse differs from the light form at four positions in the MC1R protein. Interestingly, these dark pocket mice have exactly the same change at position 230 as the arctic skua. Thus, not only is the same gene involved in body color evolution in certain species of birds, reptiles, and mammals, in some cases the same changes have arisen in the *MC1R* genes of different kinds of animals.

A second example of such precise repetition involves the jaguarundi and the golden-headed lion tamarin. Dark forms of the jaguarundi have a twenty-four-base deletion in their *MC1R* gene; the same deletion is present in the golden-headed lion tamarin, whose main body is all black, unlike other lion tamarins (plate R).

The evolution of rumination in mammals, galactose use in yeast, albinism in cave fish, and dark body color in various birds, reptiles, and mammals all demonstrate the reproducibility of evolution at the fundamental level of individual genes.

While in the many examples of convergent evolution in the opsins I described, the repetition of evolution was very often exact to the very same base pair, the changes in the genes involved in these other traits usually are not quite so precise. Detailed biochemical studies of ribonucleases and the MC1R receptor indicate that there are many different sites in each protein that can be changed to produce similar functional properties.

The contrast between the exact repetition of opsin evolution and the less precise repetition of evolution in other proteins underscores the fact that there may be multiple solutions to some evolutionary "problems" (adaptations), and only one or a few solutions to others. The structure of opsins is such that the exact amino acid in just a few key sites tunes the wavelengths of the visual pigment. Selection is most effective upon these sites. The structure and activity of the ribonuclease and MC1R are more accommodating and there are many different ways to change each of their activities. In other words, for some genes and traits the DNA code does not have to change in precisely the same way to have the same biological effect.

It also turns out that, for some traits, evolutionary convergence can also arise entirely from different genetic starting points.

Different Means to Similar Ends

One of the key inventions of Antarctic fish was antifreeze, which is made up of proteins with an unusual repeating structure of just three amino acids, usually threonine-alanine-alanine or threonine-proline-alanine. The core repeat arose from part of the code of a digestive enzyme that contained the sequence. The ancestry of the antifreeze could be traced to the enzyme gene by virtue of its noncoding sequences. Because the text immediately adjacent to the antifreeze gene is so strikingly similar to that at the enzyme gene, it is clear that antifreeze arose from a piece of DNA encoding part of the enzyme as well as some neighboring code.

Arctic fish also thrive in very cold water and have antifreeze in their bloodstream and tissues. The Arctic antifreeze proteins are also made up of repeating sequences of threonine-alanine-alanine or threonine-proline-alanine. Surely, the simplest explanation would be that a common ancestor of Antarctic and Arctic fish invented this protein and each group of species inherited their antifreezes from this ancestor.

But, in this case, the similar appearances of the antifreezes are deceiving.

The Arctic fish antifreeze evolved in a different way and at a different time from the Antarctic fish antifreeze. We know this from many pieces of evidence. First, in the fish evolutionary tree, the Antarctic and Arctic groups are distant from each other and belong to different orders. Second, the freezing of the North Atlantic and North Pacific oceans occurred much more recently, about 2.5 million years ago, indicating that the driving force for evolution of antifreeze was separate from that in the Southern Ocean, which dropped to freezing temperatures about 10 to 14 million years ago. Of course, the latter fact does not rule out the possibility that some Antarctic fish could have migrated north and given rise to Arctic types. But we can eliminate that scenario based upon the traces of antifreeze origins in DNA.

Two key clues reveal that the Arctic antifreeze had a separate origin. First, there is no trace of any similarity to the digestive enzyme gene that was the source of the code for the Antarctic antifreeze. Second, and more decisive, the two antifreezes are produced by entirely different processes. In the Antarctic fish, repeats of threonine-alanine-alanine or threonine-proline-alanine are encoded in multiple tracts, which are separated by the sequence leucine-isoleucine-phenylalanine. These spacers are positions where the protein is processed into smaller antifreeze peptides. In the Arctic fish, the spacers have an entirely different sequence and are processed by a different enzyme. So, while the antifreeze peptides are incredibly similar, they are produced from proteins with different internal spacers that must have had different origins. The antifreezes are analogues, not homologues.

The explanation for the remarkable convergence in the sequence of the Arctic and Antarctic antifreeze peptides must be natural selection for their performance in preventing ice formation in the animal. Extensive biochemical studies have shown that the antifreeze peptides work by adsorbing to ice crystals and preventing their growth. These peptides are linked to carbohydrates via their threonines, and the carbohydrates play a critical role in the interaction with ice crystals. The simple repeat threonine-alanine-alanine appears to be best suited to forming a repeating structure that can interact with the regular repeating structure of ice crystals. The evolutionary convergence of the Arctic and Antarctic antifreezes illustrates that there was more than one way to generate this repetitive code and a functional antifreeze.

The remarkable similarity of the antifreezes, given their different origins, raises the question of whether it is generally true that to have similar functions, molecules must have similar sequences.

To answer that riddle, I am going to propose a little game. Below are the sequences of four small proteins found in nature. Look carefully at their sequences (shown using the single-letter abbreviations for the twenty amino acids).

1. VCRDWFKETACRHAKSLGNCRTSQKYRANCAKTCELC
2. ZFTNVSCTTSKECWSVCQRLHNTSRGKCMNKKCRCYS
3. CRIONQKCFQHLDDCCSRKCNRFNKCG
4. ZPLRKLCILHRNPGRCYQKIPAFYYNGKKKQCEGFTWSG
   GCGGNSNRFKTIEECRRTCIRKD

Do any similarities pop out at you?

No?

Well, don't feel bad. I don't see any either, but they do have something in common.

Here is a clue to the story. The fourth protein comes from a snake. I have been interested in snakes all my life and I have always taken the opportunity to look for snakes when I travel to places with interesting species. This protein sequence comes from the only snake that ever truly terrified me. I visited a very small reptile collection near Lake Baringo in Kenya and one of the keepers was delighted to bring out a very jumpy nine- or ten-foot-long black mamba to show me. The snake was so fast and agile. I kept trying to back farther away, but the handler just brought him closer.

Had the handler made a mistake, neither he nor I would have had long to think about it. A black mamba bite can be fatal in thirty minutes. The venom contains potent neurotoxins (the sequence of a major toxin is the fourth protein sequence in the list above). The neurotoxin kills by blocking so-called potassium channels. These channels play a critical role in the electrical signals that pass among neurons and muscles. When their function is blocked, so are nerve and muscle function. Victims of black mamba bites typically display neurological and muscle problems and, if untreated, die of respiratory paralysis.

The other three proteins in the list are also potassium channel blockers found in venom. Here is the amazing part of the story—the first protein comes from a sea anemone, the second protein is from a scorpion, and the third protein is from a marine cone shell snail. Each of these animals, as well as the black mamba, belongs to a different animal phylum—the anemone is a cnidarian, the scorpion is an arthropod, the cone shell is a mollusk, and the snake a vertebrate. Each of these venom toxins evolved independently of one another and represent different molecular solutions for the binding to and blocking of potassium channels of their prey. They are different molecules with different origins but a common, deadly purpose.

I HOPE that you are more than just a little impressed, perhaps even dazzled, by the evolution of these venom toxins, as well as by the other examples of evolution repeating itself in this and earlier chapters. I believe that they are among the most powerful and important demonstrations of how evolution works in nature. They derive their special significance from two elements—repetition and detail.

There is a variation of an old Latin proverb that states "*repetitio est mater doctrinae*"—repetition is the mother of learning. What is true for education holds true for science. While individual examples of species adaptations are instructive, the repetition of events, sometimes in precise detail, teaches us that when similar forces converge, similar results emerge. Evolution is remarkably reproducible.

Thus far, my descriptions of evolution repeating itself have focused on the "how"—on the steps involved in the making of similar adaptations, or in the loss of traits. There remains the question of "why"—why is it that evolution can and does repeat itself? The answers to that question boil down to the three key ingredients—chance, selection, and time—and the everyday math we first learned in chapter 2. That arithmetic might be fuzzy at this point (or may have been fuzzy from the start), but the interplay of these three ingredients and their impact on the text of DNA reveal exactly why the same events can and do happen again and again.

Chance, Necessity, and the (Re)Making of the Fittest

In chapter 2, when the key ingredients of evolution were introduced, we did not yet have the benefit of seeing the steps of evolution in DNA. Now we have seen that the evolution of opsins, ribonuclease, MC1R, galactose enzymes, etc., has involved repeated and sometimes identical changes in the text of their genes. What was "just" theory in chapter 2 has now been documented by the study of species at their most fundamental level—the individual elements of the text of DNA.

The meaning of the discoveries in this chapter can be encapsulated as a series of general statements about the ingredients of evolution:

i. Given sufficient *time*,

ii. identical or equivalent mutations will arise repeatedly by *chance*, and

iii. their fate (preservation or elimination) will be determined by the conditions of *selection* upon the traits they affect.

I will spend the remainder of this chapter focusing on these statements, using the real arithmetic of mutation, the real-world biology of species, and real examples from this and past chapters to demonstrate why evolution can and does repeat itself. The calculations and the facts obtained from the DNA record show, beyond any doubt, that the combination of random mutation and natural selection, over time, easily accounts for biological evolution.

The proof will come from crunching some fairly large numbers. I will forewarn you that at some point along the way the thought "that's impossible" might occur to you. In fact, it is a common tactic for deniers of evolution to conjure up some purported mathematical analysis against the probability of Darwinian evolution. These arguments *always* omit one or more important factors. We shall see that when all factors are considered, evolution via specific, selected changes in DNA is not merely probable, it is abundantly so.

Chance: "identical or equivalent mutations will arise repeatedly by chance"

Let's begin with some hard evidence from the evolution of ultraviolet vision in birds. In four different orders, there are both ultraviolet-sensing and violet-sensing species. This means that the switch between violet-sensing and UV-sensing capabilities must have evolved at least four separate times. The difference between birds is always correlated with a particular amino acid, at position 90 in their short wavelength (SWS) opsin; birds with a serine in this position are tuned to violet, birds with a cysteine here are tuned to UV.

This amino acid is encoded by DNA positions 268–270 in the text of the birds' SWS opsin genes. Close scrutiny of the DNA text of the birds' SWS opsin gene reveals that the difference between serine and cysteine involves just a single letter of the DNA text at position 268, shown in table 6.1.

The zebra finch, herring gull, rhea, and budgerigar each belong to different orders. The key difference in their opsins involves a mutation from A to T at position 268 that must have occurred four times.

How likely is it that the exact same mutation would occur in different species? Time for some arithmetic.

The per site rate of mutation averages about 1 per 500,000,000 bases in DNA in most animals—from fish to humans. This means that the

Table 6.1. Repeated evolution of a UV-sensitive opsin

Species	DNA sequence	Amino acid	Violet or UV?
Zebra finch	TGC	cysteine	UV
Duck	AGC	serine	Violet
Herring gull	TGC	cysteine	UV
Razorbill	AGC	serine	Violet
Ostrich	AGC	serine	Violet
Rhea	TGC	cysteine	UV
Budgerigar	TGC	cysteine	UV

exact A at position 268 in one copy of the bird SWS opsin gene will be mutated, on average, about once in every 500 million offspring. It has two copies of the gene, so this cuts the average* to 1 in 250,000,000 chicks. However, there are three possible kinds of mutations at this site: A to T, A to C, and A to G. Based on the genetic code, only the A to T mutation will create a UV-shifting cysteine. If the probability of each mutation is similar (they aren't, but we can ignore the small difference), then one out of three mutations at this position will cause the switch. One A to T mutation will occur in roughly 750,000,000 birds (that's 750 *million*).

Seems like a long shot?

Not really. It is important to factor in the number of offspring produced per year. According to long-term population surveys, many species consist of 1 million to more than 20 million individuals. With annual reproduction, a plentiful species like a herring gull will produce at least 1 million offspring in a year (probably a very conservative number). Divide this into the rate of one mutation per 750 million birds; the result is that the serine-to-cysteine switch will arise once every 750 years. That may seem like a long time in human terms but we need to think on a much longer timescale. In 15,000 years, a

*I am simplifying the math by using "averages," not formal probability.

short span, the mutation will have occurred 20 separate times in this species alone.

The four orders that these birds belong to are ancient—their ancestors have had tens of millions of years to evolve UV or violet vision. At the rate calculated in gulls, the A to T mutation will occur more than 1200 times in 1 million years in just this one species. Getting the idea?

Now, what if an evolutionary change need not be so exact? I explained that different mutations in *MC1R* are responsible for dark snow geese, arctic skuas, bananaquits, and the dark color in a handful of other species (I am sure that the number of animal species with *MC1R*-affected variants is very large, but I am talking here only about the small number studied by biologists so far). From what we know already, it is clear that there must be at least ten different ways to mutate *MC1R* to darken fur, plumage, or scales. With ten target sites in the gene, and the same mutation rate (the same because the text of all genes is equally mutable), what are the probabilities that an *MC1R*-caused black variant will arise? It is 10 times higher than the probability of the exact change in the SWS opsin. Therefore, 1 in about 75 million offspring will be black. The frequency of black variants of a species per unit of time will depend on the rate of offspring production. A species that produces 750,000 offspring in a year will produce a new black variant every 100 years (10,000 new black variants in 1 million years). A species that produces 7.5 million offspring will produce 1 new black variant every 10 years. Even a species that produces only 75,000 offspring a year will produce 1 black variant every 1000 years.

Is it now so surprising that black mice, black birds, and dark lizards have mutations in the same gene? Or that some species have the exact same change in the *MC1R* gene?

What about fossil genes? Are they hard or easy to evolve? They are *very* easy to evolve. While there are usually relatively few ways to change a gene in a new functional way, there are *lots* of ways to cripple a gene. About 5 percent of all single-letter changes will disrupt a gene. In addition to these simple "typos," any insertion or deletion of text that is not a multiple of three letters will disrupt the text of a gene. Small insertions or deletions are quite common. Based upon these rates, it is probably 50 to 100 times "easier" (i.e., more likely) to disrupt a gene than it is to make a precise specific single mutation. Applying the same math as before, 1 in about 2 million animals will be born with a new potential fossil gene. We can see how often fossil genes arise, as well as far more precise mutations, as a function of reproductive rates in table 6.2.

**Table 6.2. How often will a similar mutation arise
in one gene in one million years?**

Reproductive rate (per year)	Number of Equivalent Sites		
	1	10	100
12,000	10	100	1,000
120,000	100	1,000	10,000
1,200,000	1,000	10,000	100,000

With this table as a reference, now consider that there are an estimated 10,000 bird species on Earth today. Looking at the numbers, it is abundantly clear that the same mutations arise repeatedly in all but the most rare living species and have occurred uncountable numbers of times in their extinct ancestors.

This picture is not unique to bird evolution. Many other groups of animals have population sizes and reproductive rates in these ranges, and many exceed them by a great margin. We need not do the calculations for the enormous populations of certain fish, insects, crustaceans, etc., to know that the same mutations reoccur very frequently.

But while mutations have been and are plentiful, whether or not a new, potentially "useful" mutation is preserved or lost is another matter—a matter of not getting lost by chance in the first few generations, and then a matter of selection.

**Selection: "their fate . . . will be determined by the
conditions of selection upon the traits they affect"**

In the past four chapters, we have seen how DNA information is
preserved, expanded, modified, or destroyed by the action, or inac-
tion, of natural selection. I have described the different fates of the
text of DNA under three different conditions of natural selection. In
chapter 3, we saw the power of purifying selection to preserve text for
eons in the face of the steady bombardment of mutations. In chapter
4, we saw how positive natural selection on duplicated genes and for
fine-tuning changes in genes made new information and traits out of
old text. In chapter 5, we witnessed how, in the absence of natural
selection to preserve genes, DNA text decays and erodes away. And
here in chapter 6, we have seen how the same equivalent changes occur
in DNA and can be selected for (or allowed, if selection is relaxed)
again and again.

At the fundamental level of DNA, selection affects the relative success of alternative forms of individual genes. Given two sequences, A
and B, that differ by one or more letters of code, there are three possible fates of A and B, depending upon the conditions of selection. If
sequence A affects survival or reproductive success in a superior way
to sequence B, A will be favored. Conversely, if B improves survival or
reproduction over A, B will be favored. Two other possibilities are that
A and B perform identically, or affect a trait that is no longer necessary for survival or reproduction. In these cases, the frequency of A
and B will "drift" according to random fluctuations in the numbers of
individuals with each form.

For any new mutation, then, there are three possible fates. It may be
actively preserved, actively rejected, or may be neutral and ignored by
selection. If a bird, for example, has an SWS opsin gene with AGC at
positions 268–270, it usually sees in the violet range. Now consider the
nine ways this sequence could change by mutations occurring at each
of the three bases:

A Juvenile icefish. The transparent appearance is due to evolutionary loss of scales and red blood cells. (Photograph by Flip Micklin.)

B Adult mackeral icefish, *Champsocephalus gunnari.*

C Color polymorphism in Northwestern garter snake.

(Photographs by Jeremiah Easter.)

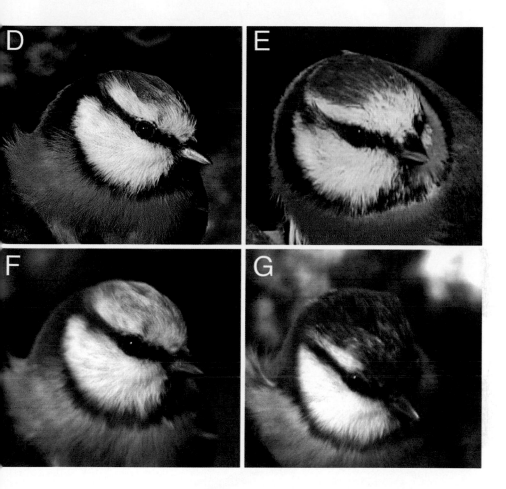

D-G UV-reflecting feathers in the blue tit.

D, F Male blue tit as seen in visible (D) and ultraviolet (F) light.
Note that the blue crown patch is highly reflective in the UV.
(Photographs courtesy of Staffan Andersson, University of Goteborg, Sweden.)

E, G Application of UV-filtering sunscreen has no effect on blue patch
in visible light (E) but blocks UV reflectance of the blue crown (G),
and reduces male mating success. (Photographs courtesy of Staffan Andersson,
University of Goteborg, Sweden.)

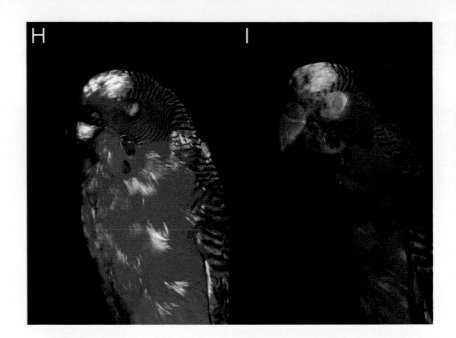

H, I The budgerigar (*Melopsittacus undulatus*) is brightly colored (H) and the forehead and cheek feathers reflect in the ultraviolet light (I), which affects mating success.
(Photographs provided by Walter Boles, © Australian Museum.)

J, K Blue-winged mountain tanagers (*Anisognathus flavinuchus cyanoptera*) in visible (J) and ultraviolet (K) light. The bird has very strong UV-reflecting feathers on its wing.
(Photograph courtesy of Rob Bleiweiss, University of Wisconsin. From R. Bleiweiss, *Proceedings of the National Academy of Sciences, USA* 101 (2004): 16561–64, © 2004 National Academy of Sciences, USA.)

L The douc langur, *Pygathrix nemaeus nemaeus*. This monkey has evolved a specialized gut for ruminating leaves. (Photograph © Paul Bratescu.)

M The white form of the lesser snow goose. (Photograph by Frans Lanting.)

N The melanic "blue" form of the lesser snow goose. The color difference between the blue and white forms is determined by the *MC1R* gene. (Photograph by Thomas Mangelson.)

O The arctic skua is found in both melanic (left) and lighter forms (right). The difference between the two forms is determined by the *MC1R* gene. (Photograph by Torbjörn Persson.)

P The black form of the fairy wren, *Melurus leucopterus*.
(Photograph by Melanie Rathburn.)

Q The blue form of the fairy wren, *Melurus leucopterus*. The color
difference between the blue and black forms is determined by the
MC1R gene. (Photograph by Melanie Rathburn.)

R The dark form of the lion tamarin. The black body coat versus the orange form is determined by the *MC1R* gene. (Photograph by Claus Meyer.)

Original	AGC ⟶	serine	violet vision

Mutations at base 270	AGT ⟶	serine	violet vision
	AGA ⟶	arginine	?
	AGG ⟶	arginine	?

Mutations at base 269	ACC ⟶	threonine	?
	ATC ⟶	isoleucine	?
	AAC ⟶	asparagine	?

Mutations at base 268	TGC ⟶	cysteine	UV vision
	GGC ⟶	glycine	?
	CGC ⟶	arginine	?

If selection was not operating, we would expect to see all of these variations in bird opsins. However, a survey of 45 species from 35 families revealed that *all* species had either serine or cysteine encoded at this position. The probability of this pattern occurring at random is infinitesimally small. This means that all of the other possible changes at this position have been consistently rejected again and again during the evolution of birds. Such is the power of natural selection.

While simple statistics can signal nonrandom patterns in DNA, we are not limited to mathematics in testing whether selection is operating in a particular way. Experiments in the laboratory and the physiology of species give additional information that, together with the DNA text, provides the full picture. In this instance, birds with cysteine at this position have been demonstrated to possess and use UV vision, while species with serine at this position have an opsin that does not detect UV wavelengths. The only explanation for the selective occurrence of cysteine and serine at this position is natural and sexual selection. And the best explanation for the repeated evolution of cysteine at this position is that similar conditions, favoring the capability of UV vision, have existed and do exist in different orders, families, and species of birds.

Likewise, the convergent evolution of trichromatic vision in howler monkeys, of ribonucleases in ruminants, dark pigmentation in birds, mammals, and reptiles, antifreezes in cold-water fish, and potent neurotoxins are best explained by similarities in the conditions of selection that favored the emergence of similar traits.

While the advantages of antifreezes and of prey-killing toxins are obvious, their evolution demonstrates how selection acts on the materials that are available. It has fashioned two nearly identical antifreezes from two entirely different pieces of code, as well as deadly toxins from many different starting materials. In these examples, necessity was clearly the mother of invention, but it was the combination of random mutation and selection that forged these inventions.

If, however, a trait becomes unnecessary through some shift in lifestyle, selection will be blind to the genes that affect it. Mutations that disrupt the gene are inevitable and may occur anywhere in the text. We have seen that the SWS opsin has been disrupted at least five different times, in five different ways, in certain vertebrates and that the seven genes of the galactose pathway have been disrupted at least three times in yeast. Again, similar conditions (dispensability) have led to similar outcomes.

Not every time a potentially "useful" mutation arises will it spread completely through a species. In fact, most of the time, new mutations are lost by chance before they reach a significant frequency. Only a fraction of new mutations, depending upon the magnitude of the advantage they confer, will be spread by selection. The figures in table 6.2 on the recurrence of mutations demonstrate that the frequency of mutations provides ample opportunities for evolution, even if those opportunities are seized only occasionally.

It is very important to stress that because species often occupy multiple habitats, a mutation may be favored in some locales and selected against or ignored in others. This results in species, like the rock pocket mice, jaguar, or snow geese, being variable in many traits across their range. While I have described how frequently new mutations arise, it is usually the case that two or more alternative forms of a gene are present at substantial frequencies in a population or species. Evolution is then not a matter of "waiting" for a new mutation, but *the preferential increase or decrease of alternative forms in response to changes in conditions*. As we saw in chapter 2, the timescale of selection is such that, once established, particular traits can spread quickly or disappear rapidly.

ALMOST TWO THOUSAND years ago, the writer-philosopher Plutarch came very close to describing the nature of evolution, and the likelihood of repetition, in his *Life of Sertorius* (quoted at the front of the chapter). He astutely emphasized the length of time available ("the long process of time"), chance ("fortune"), and the large number and diversity of material available ("the number and variety of subjects to be wrought upon be infinite"), and concluded that events should be repeated in history ("numerous coincidences should spontaneously occur" and "effect this similarity of results").

Plutarch aptly described the probabilistic dimension of history, but not the deterministic part of the evolutionary process—that is, selection. Out of the vast numbers of possible events that do occur in the random scrambling of DNA, selection weeds out the majority and favors but a few. The making of the fittest is not merely a matter of chance but, as the great biologist Jacques Monod put it over thirty years ago, it is a matter of *chance and necessity*. The repeated evolution of traits is a product of both forces, the probability of equivalent mutations and the similarity of the conditions of selection.

We have now come a long way since the theoretical arithmetic of chapter 2. However, you may be wondering "this is all well and good for mice and yeast and parakeets, but what about humans?"

The interplay of chance and necessity is not restricted to history nor to "lower species." We can see mutation and selection occurring in real time in our own species. That will be the focus of the next chapter.

The very toxic Oregon rough-skinned newt. *Photograph by Steven Holt.*

Chapter 7

Our Flesh and Blood: Arms Races, the Human Race, and Natural Selection

. . . .

The mortal enemies of man are not his fellows of another continent or race; they are the aspects of the physical world which limit or challenge his control, the disease germs that attack him and his domesticated plants and animals, and the insects that carry many of these germs. . . .
—W. C. Allee, *The Social Life of Insects* (1939)

IT IS RUMORED THAT THE MOST COMMON LAST words humans utter are "Hey, hold my beer and watch this!"

For one unfortunate Oregon man in 1979, they probably were. After a day of heavy drinking, and prompted by a buddy's dare, this otherwise healthy twenty-nine-year old swallowed an eight-inch-long newt. Within ten minutes his lips tingled. Then he began feeling numb and weak, and told his friends he thought he was going to die. He refused to go to a hospital and then promptly went into cardiac arrest. Although his heart was restarted, all other medical efforts failed and he died in less than twenty-four hours. For being the first and only human known to die from newt poisoning, and for voluntarily removing his genes from the human gene pool, I nominate this pioneering gourmand for a Darwin Award.

Docile, soft, and downright cute, newts do not inspire much fear. But the Oregon rough-skinned newt (*Taricha granulosa*) packs one hell of a wallop. Its skin is loaded with a toxin called tetrodotoxin (TTX for short), enough to kill several humans. TTX is a potent blocker of another type of ion channel, sodium channels, that are critical for nerve function. TTX causes paralysis, respiratory distress, abnormal heartbeats, and in modest doses, death. The toxin is better known from the practice of eating puffer fish in Asia, which also contain high levels of TTX. For $400 or more per person, a licensed, "expert" chef will prepare this delicacy in such a way that the diner can savor the supposedly wonderful flavor of the flesh without ingesting any of the toxin. Nonetheless, from 1974 to 1983 there were 646 reported cases of puffer fish poisoning in Japan, 179 of them fatal. I've traveled to Japan—my advice is to stick to the teriyaki and tempura.

Why would a newt weighing half an ounce pack enough poison to kill a 170-pound human? The Oregon man was a victim of, in addition to his own stupidity, an evolutionary "arms race." The newt is locked into a life-or-death battle with the common garter snake, *Thamnophis sirtalis*. While newts can escape unharmed after being swallowed by bullfrogs and other animals, the only predator known to be resistant to the newt's TTX is the garter snake.

The level of resistance varies among individual snakes, as does the level of TTX production among individual newts. And different populations of snakes and newts exhibit different degrees of resistance and toxicity, respectively. The variation in predator and prey sets the stage for a "coevolutionary" arms race in which selection favors increasingly resistant snakes and increasingly toxic newts. The escalation of the arms race through the evolutionary process results in preposterously toxic newts that easily kill other predators, and highly resistant snakes that have survived newt dinners.

Arms races are examples of evolution in a fast-forward mode. The intense selection on each party accelerates the rate of evolutionary change such that biologists can catch species in the process of evolving.

All snakes that try to eat a newt exhibit some signs of TTX poisoning: they cannot hold their heads steady, and they become flaccid and unable to right themselves. In laboratory studies, most snakes release the newt and recover. But even the largest snakes that manage to swallow a whole animal can be paralyzed and succumb to the poison. Only a fraction of snakes are able to successfully consume a newt.

You must be wondering, why do the snakes even go after such a difficult and dangerous meal? The answer is familiar to all of us who have overeaten a rich meal. There is a trade-off between taking advantage of an abundant food source and feeling lousy for a while. Those snakes with greater resistance can feed on newts, others cannot. While the snakes may get woozy, that beats not eating at all. And, because resistance is inherited, their young will also have this "advantage" over less resistant snakes. Maybe an analogy is the marshmallow and Jell-O salad at a picnic. It makes me throw up, but the rare aficionado gets to feast on as much as he or she wants.

If newts or puffer fish were important items on our menu, we, too, might be in an arms race with them. Unfortunately for the Oregon newt swallower, we are not. However, we humans have been and are engaged in arms races with deadly foes. Some of these battles have left very strong marks on the genetic makeup of populations of our species. And some battles are yet to be won.

In this chapter, I will focus on seeing the evolutionary process at work in humans—in our own flesh, blood, and DNA. I will first illustrate how our physical environment affects human evolution and our genetic makeup. Then I will discuss how arms races with infectious agents have shaped human history and human genes. We are the descendants of the fittest who have survived malaria, plague, poxes, and scores of other scourges, and some of our genes show the scars from these battles. Finally, I will explain how cancers arise in our bodies through a type of evolutionary process, and how knowledge of that process is leading to the design of new weapons to battle the disease. Throughout the examples in this chapter, we will see that the process of selection operates everywhere there is variation and competition in life. We will also see that mutation and selection are hardly mere "theory"—they have been and continue to be a matter of life and death.

Man Against the Sun

One of the most obvious characteristics that differentiates people from the various parts of the world is skin color. Are color differences a product of evolution, brought about by selection? Or, alternatively, is skin color "just" a reflection of relatedness among people found in the same region?

These questions have been contemplated for a very long time. In fact, ideas about human skin color and fitness predate Darwin. Although Darwin was unaware of it at the time of *On the Origin of Species*, William Charles Wells, an American-born physician, had some forty years earlier articulated the essence of the principle of natural selection in remarkably clear terms. Furthermore, he connected it to some of the forces affecting human diversity.

Wells was born in South Carolina to Scottish immigrants in 1757. He was sent to Scotland for schooling and at age thirteen entered the University of Edinburgh, but then soon returned in 1771 to Charleston to apprentice under a Dr. Alexander, a distinguished botany enthusiast and devotee of Carl von Linné. After completing his medical training, Wells struggled as a physician, but he achieved exceptional success with various scientific contributions ranging from studies of how muscles contract to the treatment of vision problems. He also made the first correct explanation of how dew is formed, for which he won the Rumford Medal of the Royal Society.

In 1813, he presented a paper entitled (in the style and language of the day) "An Account of a Female of the White Race of Mankind, Part of Whose Skin Resembles That of a Negro." This paper was finally published in 1818, after his death and as part of a longer body of work, including Wells's autobiography (figure 7.1). Wells did not merely report his findings, but made "Some Observations on the Causes of the Differences in Colour and Form Between the White and Negro Races of Men."

Wells noted that "negroes and mulattos" enjoy immunity from certain tropical diseases. He also observed that all animals tend to vary in some degree and that breeders improve their domesticated animals by selection. He then made the same leap that Darwin would make forty years later, noting that what breeders do

> by art, seems to be done with equal efficacy, though more slowly, by nature, in the formation of varieties of mankind, fitted for the country which they inhabit. Of the accidental varieties of man, which would occur among the first few and scattered inhabitants of the middle regions of Africa, some one would be better fitted than others to bear the diseases of the country. This race would consequently multiply, while the others would decrease; not only from their inability to sustain the attacks of the disease, but from their incapacity of contending with their more vigorous neighbours. The colour of the vigorous race I take for granted, from what has been already said, would be dark. But the same disposition to form varieties still existing, a darker and a darker race would in the course of time occur; and as the darkest would be the best fitted for the climate, this would at length become the most prevalent, if not the only race, in the particular country in which it had originated.

Darwin learned of Wells's work a few years after publication of *On the Origin of Species*. In later editions, Darwin reviewed the many works concerning evolution that preceded *The Origin* in "An Historical Sketch." In the fourth edition of *On the Origin of Species*,

TWO ESSAYS:

ONE

UPON SINGLE VISION WITH TWO EYES;

THE OTHER

ON DEW.

A LETTER

TO THE

RIGHT HON. LLOYD, LORD KENYON

AND

AN ACCOUNT

OF

A FEMALE OF THE WHITE RACE OF MANKIND,
PART OF WHOSE SKIN RESEMBLES THAT OF A NEGRO;

WITH

SOME OBSERVATIONS ON THE CAUSES OF THE DIFFERENCES IN
COLOUR AND FORM BETWEEN THE WHITE AND NEGRO
RACES OF MEN.

BY THE LATE WILLIAM CHARLES WELLS,
M.D. F.R.S. L. & E.

WITH

A MEMOIR OF HIS LIFE,
WRITTEN BY HIMSELF.

LONDON:
PRINTED FOR ARCHIBALD CONSTABLE AND CO. EDINBURGH,
LONGMAN, HURST, REES, ORME, AND BROWN,
AND HURST, ROBINSON, AND CO. LONDON.

1818.

FIG. 7.1. **Frontispiece of William C. Wells pamphlet published posthumously in 1818.** Wells articulated the first accurate description of the concept of natural selection. *Image courtesy of the University of Wisconsin Library.*

in 1866, Darwin credited Wells as having "distinctly" recognized "the principles of natural selection," and "this is the first recognition which has been indicated." Darwin was right to acknowledge Wells, but was Wells right about selection on skin color? In order to evaluate Wells's hypothesis, we have to consider the physiology and genetics of skin color and to scrutinize the genes of different ethnic groups.

The lightness or darkness of skin (and hair) is primarily a matter of the relative amount of a pigment called melanin that is produced by specialized cells in the skin called melanocytes. The process of melanin formation is fairly well understood in biochemical terms, and a central component of this process is already familiar, the melanocortin-1 receptor (MC1R), the same receptor involved in fur, plumage, and scale color. Melanin production is controlled by a hormone made in one part of the pituitary gland, called α-melanocyte-stimulating hormone (αMSH), which binds to the MC1 receptor of melanocytes and stimulates the production of melanin. The melanin is then transferred from the specialized melanocytes to skin cells and hair.

A second pathway of melanin synthesis is induced by ultraviolet irradiation. UV light induces the synthesis of MC1R and αMSH, which in turn leads to the production of increased levels of melanin. This is the basis of tanning in response to sunlight.

The melanin pigment is a natural sunscreen. It is very efficient at absorbing the various wavelengths of sunlight. Because of its structure, it absorbs UV radiation. UV light causes damage to cells, and worse, it acts directly on the bases of the DNA code and induces alterations that lead to permanent changes in the text—UV is a potent *mutagen*. Chemicals used in commercial sunscreens are selected for their UV-screening properties. It is also important to know that not all UV radiation is bad. UV is essential to induce the production of vitamin D_3 in the skin, which plays a critical role in calcium absorption, which in turn is important for bone formation and maintenance. Vitamin D deficiency causes osteoporosis and, when severe, rickets. This is why milk is fortified with vitamin D.

The amount of UV radiation that hits the Earth varies from region to region. A number of factors determine what kinds of UV radiation are experienced by humans, and how intense it is. These include the length of the rays' path through the atmosphere (which is affected by season, the time of day, and latitude), the altitude, gases in the air, and the amount of reflection on surfaces (snow, water, etc.). The UV content of sunlight is greatest at latitudes within 30 degrees of the equator. In the winter in Boston, for example, there is not enough sunlight to trigger vitamin D production in the skin.

Recently Nina Jablonski and George Chaplin of the California Academy of Sciences found a close correlation between the levels of UV radiation in different regions of the world and variations in skin pigmentation. Is this variation in our physical environment responsible for differences in skin color? If so, how can we tell that skin color has been under selection?

The most logical places to start looking for answers are the *MC1R* genes of fair- and dark-skinned individuals. Fair-skinned individuals of northern Europe, such as the Scots or Irish, often have red hair and freckles, and their skin is much more sensitive to sunlight than darker-skinned individuals. By studying the inheritance of skin and hair color in families, it is clear that variation in the *MC1R* gene is linked in humans, as in other animals, to differences in skin and hair color. Red hair in Europeans, or those of European ancestry, is almost entirely accounted for by a number of different, specific mutations in the *MC1R* gene that cause the replacement of one single amino acid with another amino acid. In European and Asian populations, there are at least thirteen different variants of the *MC1R* gene; ten of these alter the MC1R protein and three do not (they are synonymous substitutions).

In contrast, in Africans, while there are five variants of the *MC1R* gene, they are all synonymous. There are no variants of the MC1R protein. The different ratios of nonsynonymous to synonymous substitutions in *MC1R* in non-Africans (10:3) and Africans (0:5) are, statistically speaking, highly significant and could not be explained by random chance alone. Rather, something has prevented the occurrence of changes in the MC1R protein in Africans—that, of course, is natural selection. Mutations do occur at the *MC1R* gene—we know that and can confirm it by the existence of the five synonymous variants. But the great constraint on *MC1R* variation in Africans suggests that selection has maintained high levels of melanin production. This makes perfect sense in terms of the protective benefit of melanin in zones of high sunlight and UV radiation.

Wells was right about dark skin color being "better fitted" to Africans.

Because Europeans are so variable, and lighter-skinned, we might also ask whether this is due to selection for light skin or the relaxation of selection in melanin production. At the moment, there are good arguments for both explanations. At northern latitudes, selection for melanin production may be relaxed. But, because some UV absorption is needed to stimulate vitamin D production, it is possible that lighter skin is an adaptation to lower levels of sunlight. No matter which is the case, the evolution of human skin color and the *MC1R* gene demonstrates that as humans spread out across the globe, the conditions of selection varied in different regions, the amount and quality of sunlight being just one obvious variable.

Other critical factors in human history are the different disease-causing organisms encountered in particular regions. Wells speculated that since dark-skinned individuals were immune to some diseases, dark skin might be the basis for that advantage. We'll see that, although his reasoning was incorrect, his observations about resistance and his inference that the capacity to sustain the attacks of disease would be a major determinant of fitness were certainly on the right track. Infectious diseases are powerful selective agents that have left a strong mark on human evolution. We'll see that their importance also helps to explain the great paradox of why certain genetic diseases persist at high frequencies in some populations—a paradox that was solved by another physician whose contributions have been generally overlooked.

Germ Warfare

Anthony Allison was raised on a farm in the 1930s and '40s in the Kenya highlands overlooking the Great Rift Valley. Even as a young schoolboy, he noticed the great variety of plant and animal life specialized to the different habitats across Kenya, and the great diversity of indigenous tribes and their languages. A visit to Louis Leakey's famous Olduvai Gorge excavation site prompted an early interest in the question of human origins and evolution. As a teenager, he read *On the Origin of Species* and *The Descent of Man* and then at Oxford University was exposed to the mathematics of R. A. Fisher, J. B. S. Haldane, and Sewall Wright that related evolution and selection to genetics. There was, at the time of his schooling, no example of natural selection operating on a gene in humans. Early in his career, his upbringing and a series of events would inspire and enable Allison to unearth such evidence.

In 1949, in between basic science studies and beginning medical school, Allison participated in an Oxford University expedition to Mount Kenya. While his classmates focused on plants and insects, Allison collected blood samples from tribes all over Kenya for blood grouping and other tests.

One of the key tests was for the prevalence of the sickle cell trait. Discovered in 1910, the trait is named for the sicklelike appearance of red blood cells in affected individuals (figure 7.2). It was known that the disease was genetically recessive; cells of individuals carrying one copy of the sickle cell mutation (carriers) exhibited the trait under certain circumstances, while individuals carrying two copies of the mutation had the full-blown disease. That year of 1949 was a milestone in understanding the basis of sickle cell disease. The great biochemist Linus Pauling and his research group discovered that the hemoglobin molecule, the oxygen-carrying protein of our red blood cells, was abnormal in sickle cell patients.

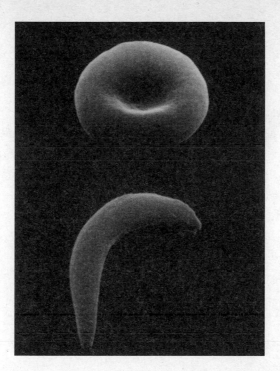

FIG. 7.2. **Normal- and sickle-shaped red blood cells.** The sickle shape is caused by abnormal hemoglobin. *Photographs courtesy of Dr. James White, University of Minnesota Medical School.*

During the 1949 expedition, Allison found that the frequency of sickle cell trait among tribes differed a great deal. In those tribes living close to Lake Victoria or the Kenya coast, it exceeded 20 percent. In tribes living in the highlands or arid regions, frequencies were less than 1 percent. Allison puzzled over these observations. Why, since sickle cell disease is bad, would the frequency of the trait be so high? And, why is it high in some tribes and not others?

He came upon a exciting and truly brilliant idea. Allison thought that maybe the sickle cell frequencies were related to resistance to malaria. He knew that malaria and the mosquitoes that carry it were much more prevalent in low-lying, moist areas, and nearly or completely absent at higher elevations or where freshwater was lacking.

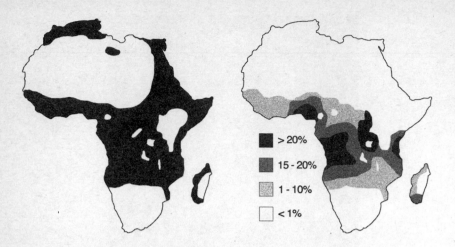

> 20%

15 - 20%

1 - 10%

< 1%

FIG. 7.3. **The geography of sickle cell hemoglobin and malaria.** These maps show the close correspondence between the frequency of the sickle cell trait (right) and the distribution of malarial parasites (left). *Maps are based on A. C. Allison (2004),* Genetics *166:1591. Redrawn by Leanne Olds.*

It would be a few years before he could test his idea, after he received his medical training. In 1953, he was able to demonstrate that patients with the sickle cell trait were relatively resistant to induced malarial infections and that children with sickle cell harbored lower numbers of parasites than children with normal hemoglobin. Allison surveyed nearly 5000 East Africans and found that high frequencies of sickle cell, up to 40 percent, were indeed restricted to malarial zones while low frequencies were found in individuals who lived in areas where no malaria occurred. Allison composed a map that shows the close correspondence between the frequency of sickle cell and malarial zones across Africa (figure 7.3). The correlation cuts across tribal and language boundaries, showing quite convincingly that malaria has had a profound effect on the genetics of human populations. The evolution of the sickle cell gene is the classic textbook example of natural selection in humans. (Curiously, most texts do not credit Allison for the idea that malaria was the selective agent—see Sources and Further Reading.)

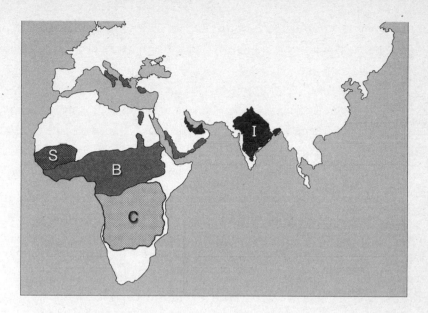

FIG. 7.4. **The human sickle cell mutation has evolved several times.**
The map shows the distribution of five occurrences of the same mutation that arose in different regions of Africa (Senegal, Benin, and the
Central African Republic), southern Europe, and the Indian subcontinent (I), and that then spread. *Map drawn by Leanne Olds.*

In the half century since Allison's pioneering studies, much more evidence has accumulated that demonstrates how malaria has left such a
strong mark on human genes. In addition to the incidence of sickle cell
disease in sub-Saharan Africa, the disease appears sporadically in
Greece and India. Sixteen percent of people living around Lake Copais
in central Greece and up to 32 percent people living in the Chalcidice
peninsula in northern Greece carry the sickle cell gene. Tribal populations in the Nilgiri region of southern India also carry the disease at a
frequency up to 30 percent. What do these regions have in common?
Until recent eradication efforts, they were hot spots of malaria.

Even more compelling evidence of the link is that the sickle cell
mutation, which is caused by a single change in the sixth triplet of the
gene (from GAG to GTG), appears to have arisen *at least five separate
times* in human populations: in Bantu People, in Benin, Senegal, and

Cameroon, and populations on the Indian subcontinent (figure 7.4). Here, again, evolution is repeating itself quite precisely, in our own species.

Under the strong selection pressure of the malarial parasite, a mutation that is generally harmful when present in two copies conveys a critical survival advantage when present in one copy. This advantage is why "bad" copies of certain genes may be abundant in human populations.

In fact, sickle cell hemoglobin is not the only example. An enzyme that I will abbreviate as G6PD plays a critical role in the metabolism of glucose and the oxidative climate inside cells. Deficiencies in this enzyme are the most common enzyme deficiencies of humans, affecting an estimated 400 million people. Thirty-four different G6PD mutations are present at high frequencies in some human populations. Want to take a guess at where those populations are?

If you guessed malarial zones, good for you! There is no way this association, with all of the frequent variants in malarial areas and none in nonmalarial zones, is due to chance. Patients with G6PD deficiency do, in fact, have lower loads of the malarial parasite than normal controls. A large study of 2000 African children showed that G6PD deficiency is associated with a reduction in risk of severe malaria of 46 to 58 percent. Furthermore, parasite growth has been shown to be inhibited in G6PD-deficient cells. Evidently, the altered oxidative environment in G6PD-deficient red blood cells compromises the life cycle of the parasite.

Other mutations help to avoid malaria by preventing the parasite from entering red cells altogether. The most severe form of malaria is caused by the species *Plasmodium falciparum*. Another species, *P. vivax*, is prevalent in west Africa. This parasite gains entry to red blood cells by latching onto a protein on the surface of red blood cells called the Duffy protein. A mutation that eliminates the expression of the Duffy protein on red cells occurs at up to a 100 percent frequency in African populations, whereas this mutation is rare or nonexistent in Caucasians or Asians. The *P. vivax* parasite cannot gain entry to the red blood cells of individuals with the mutation.

The Duffy gene mutation has clearly been favored by selection where *P. vivax* is present.

Strong selective pressure from malaria has shaped the genetic evolution of humans in many ways. But for how long? There is a fascinating archaeological record of malaria's impact on humans. The symptoms of malaria were described in ancient Chinese medical writings in *Nei Ching* (The Canon of Medicine), written around 2700 B.C. Malaria became widely recognized in Greece by the fourth century B.C. and it has been attributed to the fall of several city-state populations. It was the Romans who named the ailment *mala aria* ("bad air") because they thought it was caused by bad air, particularly that around swamps. Some archaeologists believe that malaria contributed to the decline of the Roman Empire in that expansion into malarial zones was a failure that drained the empire's resources.

Using genetics, we can get some idea of when malaria began to have an impact on human evolution. By studying genetic markers linked to mutations in the *G6PD* gene, Sarah Tishkoff of the University of Maryland and her collaborators have estimated that two particular mutations arose just in the last several thousand years. Furthermore, a study from the Harvard-Oxford Malaria Genome Diversity Project concluded that the *Plasmodium falciparum* that causes the most virulent form of the disease is also of recent origin, in the range of the last 3200 to 7700 years.

Is there any relationship between or significance to these dates?

These figures coincide with the spread of agriculture, which began roughly 10,000 years ago, and raise the possibility that malaria's effect on human evolution is relatively recent. As humans cleared forests for crops, they increased the amount of sunlit pools of water where the *Anopheles* mosquito that carries the parasite breeds. The increase in mosquito populations and the increase in human population density, including settlements around lakeshores, probably facilitated the spread of malaria and set in motion the evolutionary arms race among the parasite, mosquitoes, and humans. Today, malaria affects some 300–500 million people, and 2 million die from it every year.

Malaria is not the only disease whose prevalence has shaped the evolution of humans and may be responsible for the high incidence of particular genetic diseases. The bacterium that causes typhoid fever, *Salmonella typhi*, may similarly explain the high incidence of cystic fibrosis (CF) mutations in certain Caucasian populations. Individuals with two copies of these mutations have the disease, which was, until recently, usually fatal before the age of twenty. However, the frequency of CF mutations in populations is much higher than would be expected of a lethal-disease-causing mutation. Laboratory studies indicate that the *S. typhi* bacterium uses the CF protein to enter intestinal cells and that cells of mice bearing the most common CF mutation take up fewer *S. typhi*. So the CF mutation may allow some resistance. Frequent epidemics of typhoid fever (also called spotted fever and ship fever) throughout history may then have selected for individuals bearing the CF mutation.

All sorts of pathogens gain entry to cells via specific molecules on host cell surfaces, so mutations in cell surface molecules that confer some resistance can be expected to play a central role in the battle between pathogens and humans. It has been discovered that some individuals are resistant to HIV infection because they carry a mutation in a gene called *CCR5* that encodes part of the receptor the HIV virus uses to gain entry to cells. The emergence of HIV virus is too recent to explain the frequency of this mutation. It is most likely that the CCR5 mutation was selected for resistant to another pathogen. One proposed candidate is a virus that causes hemorrhagic fever and appears to have struck European populations in the Middle Ages.

Dodging pathogens through receptor mutations is just one facet of our arms race with germs. Our main line of defense, once infected, is our immune system, which has several means of containing, engulfing, or directly killing pathogens. In turn, pathogens have all sorts of genetic tricks for evading our immune systems, including mutations that continually alter their appearance in order to stay one step ahead of us.

We have perpetuated this arms race in modern times with our attempts to combat pathogens or their sources. Efforts to fight malaria have focused both on agents to kill the *Plasmodium* parasite and the *Anopheles* mosquitoes that carry it. A global eradication effort began in the 1950s and wiped out malaria in the southeastern United States by 1951 and in Europe by 1979. But these efforts, using DDT to kill mosquitoes and drugs to combat the parasite, faltered elsewhere and spurred evolutionary arms races with both organisms that, I am sorry to say, we have been losing.

Chloroquine, for example, is one of the safest and cheapest and, initially, was one of the most effective treatments for malarial infection. However, mutations in a single gene confer chloroquine resistance in *Plasmodium*; such resistance is now so widespread that the drug is generally useless. Resistance to mefloquine, quinine, sulfadoxine/pyremethamine, and other drugs has also evolved. Similarly, efforts to eradicate the mosquito with DDT led to the evolution of DDT resistance by mosquitoes (and catastrophic effects on animals such as raptors that were higher up on the food chain).

The situation is critical, but not hopeless. In fact, the malaria evolutionary arms race is a good example of how the application of evolutionary principles to medicine, so-called evolutionary medicine, may lift us out of the perpetual cycle of new drug/new resistance/newer drug/more resistance, etc. The key idea considers the interaction of mutation and selection. We know that with almost any drug that targets a particular parasite protein, drug-resistant mutations will evolve sooner or later. When a drug is widely used, that creates a selective condition where only resistant forms thrive and are transmitted. So, after some initial period of success, the disease comes roaring back and available drugs are ineffective.

The newer approach is to use *combinations* of drugs. The key idea is that the chance of an organism harboring resistant mutations to two or more drugs is a product of the frequency of resistance to each drug. So let's say 1 in 100 million parasites are resistant to either drug X or drug Y; then it is expected that 1 in 10,000 trillion (100 million × 100 million) parasites will be resistant to both drugs. In other words, there is a much slimmer chance of the rise of resistance to drug combinations. Combination drug therapy has been the strategy that has made all of the difference in combating the HIV virus. A new malarial combination therapy, based upon the active ingredient in a remedy first described in China in the second century B.C., hits the parasite with multiple drugs (for which resistance to each is unlikely) and appears to work well. This new regimen is, however, more expensive than chloroquine, and cost remains the main impediment to the widespread availability of effective medicines in Africa.

The modern arms race, between pesticides and pest or drugs and parasites, demonstrates that understanding the interplay of mutation and selection is no mere abstract or academic indulgence—it is a very serious and important real-world problem. Hopefully, using our brains (another product of an arms race), biotechnology (which has evolved in an explosive Darwinian fashion), and knowledge of evolutionary principles, we will eventually control, and maybe eradicate, malaria. I will close this chapter with one more vivid example of the relevance of evolution to medicine.

The Enemy Within

One of the trade-offs of being large, long-lived complex organisms, composed of trillions of cells, is that we must continuously replenish certain kinds of cells in our bodies, such as our skin, blood, and gut cells. In the production of new cells, DNA must be copied and that means sometimes mistakes are made. While mutations in all cells except our germ cells (sperm, egg) will not be passed on to the next generation, some mutations can ignite a type of evolutionary arms race within our bodies that is a leading cause of death in humans—this is cancer.

Tumors arise through a process that involves the three key ingredients of evolution in nature—chance mutations, selection, and time. The initial events in cancer formation are mutations that compromise the mechanisms that control how cells multiply and how they interact with their neighbors. Some combinations of mutations bestow selective advantages upon cells that enable them to proliferate unchecked. As these tumors grow, additional mutations occur that may give cells the ability to leave their original location and travel to, invade, and proliferate in other body tissues (metastasis).

Over the past thirty years or so, biologists have sought to understand the genetic and molecular mechanisms of cancer formation. One critical set of advances has come from identifying specific genes that are often or always mutated in particular types of cancers. A prominent example is the so-called Philadelphia chromosome, which is associated with chronic myelogenous leukemia (CML). In these cancers, breakage and attachment of chromosomes has occurred, a process that fuses one gene to another. This fusion disrupts the normal control of a potent regulatory protein called the ABL kinase. This event predisposes cells to become cancerous, and about 4400 new cases of CML occur in the United States each year.

Historically, treatment of most cancers has been rather nonspecific and has employed broadly toxic drugs and radiation. These treatments are blunt instruments that destroy rapidly dividing cells, both healthy and cancer cells, and cause many serious side effects. But the discovery of particular altered genes in various cancers raised the prospect of new therapies targeted at specific molecules involved in the disease, and a new generation of "rational" chemotherapeutic drugs are now being used to treat a variety of cancers.

One such drug has been developed that targets the ABL kinase in CML tumors and has shown great efficacy and safety. The drug, which goes by the name Gleevec or Imatinib, latches on to a particular part of the ABL kinase protein, inhibiting its activity. Gleevec is frontline therapy for CML and has produced disease remission in a large number of patients.

You have learned enough about mutation and selection that you can predict what happened as Gleevec was administered to patients. Resistance. Gleevec is essentially a toxin for CML cells, and just the garter snakes of Oregon evolve resistance to TTX or malarial parasites evolve drug resistance, some CML cells will be more resistant to Gleevec, because of additional mutations that have occurred in those cells.

Recently, Charles Sawyers and his colleagues at the Howard Hughes Medical Institute at UCLA examined how resistance evolved in CML patients treated with Gleevec. By looking directly at the ABL kinase gene in resistant patients, they found that these leukemia cells have additional mutations in the ABL kinase gene. In fact, in six different patients, the *exact same mutation occurred*—yet another demonstration of evolution repeating itself (figure 7.5). The C-to-T mutation in the ABL gene in these cancers causes the replacement of a threonine in the ABL protein with an isoleucine. Because we know the exact chemistry of how the Gleevec drug latches on to the ABL kinase, we know that this change alters the pocket that Gleevec normally can fit into so that the drug can no longer bind there and block the protein function. While the original tumor cells are controlled and killed by Gleevec, cells carrying this single mutation escape Gleevec's action, and cancer relapses.

This is, of course, bad news. But it is just the initial battle, not the whole war.

Now, knowing that Gleevec resistance will emerge in some fraction of patients, and that some of the resistance mutations will occur in the same places, new agents can be designed to circumvent the Gleevec-resistant ABL kinases. Indeed, Sawyers and collaborators at the Bristol-Myers Squibb Company have shown that another ABL kinase inhibitor (currently called BMS-354825) is active against fourteen out of fifteen different Gleevec-resistant kinases. This discovery raises the hope that patients who relapse after Gleevec treatment will have an effective second-line therapy available. The discovery also raises an additional strategy—combination therapy. Just as for malaria or HIV, by understanding mutation, selection, and evolution we have a new rational plan for beating one form of cancer.

Detailed studies of CML patients have revealed the presence of Gleevec-resistant mutations in cancer cells *before* the administration of the Gleevec drug. This is a key point. Drugs or toxins do not induce

```
Normal protein          I I T E F M T
                        -----      *      -----
Normal DNA      ATCATCACTGAGTTCATGACC
Patient 1       ATCATCACTGAGTTCATGACC
        2       ATCATCATTGAGTTCATGACC
        3       ATCATCATTGAGTTCATGACC
        4       ATCATCACTGAGTTCATGACC
        5       ATCATCATTGAGTTCATGACC
        6       ATCATCACTGAGTTCATGACC
        7       ATCATCATTGAGTTCATGACC
        8       ATCATCATTGAGTTCATGACC
        9       ATCATCATTGAGTTCATGACC
Mutant protein          I I I E F M T
```

FIG. 7.5. **The repeated evolution of a drug-resistant cancer gene.** A short section of the protein sequence (top) and corresponding DNA sequence of a gene (second line) involved in the development of chronic myeloid leukemia. In six patients (numbers 2, 3, 5, 7, 8, 9) a mutation has arisen in the same position (asterisk, C → T) that causes the tumor cells to be resistant to a drug that binds to the protein. *Based upon data of Gorre et al. (2001),* Science *293:876. Drawn by Leanne Olds.*

resistance; remember that mutation is random. What drugs do is create a selective condition in which only resistant parasites, bacteria, viruses, or, in this case, cancer cells can thrive. As a cancer grows, it becomes genetically heterogeneous because the steady beat of mutation continues. Some subpopulation of cancer cells (if the cancer is large enough) will, by chance, be drug-resistant. So, new studies are ongoing using two (and in the future perhaps a third) ABL-specific drugs in order to wipe out CML cancers before resistance can evolve. The best strategy is clearly "hit it early and hit it hard" in order to have the best chance of a complete and lasting cure.

The lessons from Gleevec are being applied to many cancers, and there is good cause for hope that cancer treatment will become increasingly tailored to the specific genetics of patients' tumors, will anticipate the evolution of resistance, and therefore will be increasingly successful.

Natural Selection:
By Any Means Necessary

I began this book with the remarkable tale of the icefish, in which the necessity of reducing the viscosity of their blood in cold water outweighed the value of having red blood cells and hemoglobin. In this chapter, we have seen that in humans, the necessity to combat the malaria unleashed by our own cultural evolution has spurred the evolution of modifications to our hemoglobin and other red blood cell proteins.

Icefish and human evolution demonstrate that natural selection works with whatever materials are available. The solutions to malaria or cold temperatures may not be the best imaginable, they were just the best available. "Bad" mutations such as the sickle cell and *G6PD* mutations and the irreversible fossilization of genes can be favored in meeting the imperatives imposed by the conditions of selection. It is all a matter of the immediate benefits outweighing the immediate costs, if only by a slim margin.

The power of these particular examples is that they run so counter to our notions of progress and design. The making of the fittest is improvisation, not a scripted process. Nature has been making it up as she goes along for more than three billion years.

Over the past five chapters, I have shown some of what is the most demonstrable evidence for evolution by natural selection at the most fundamental level of DNA. I chose the examples described in this chapter to underscore how the process of selection operates *everywhere*. Wherever there is variation—in newts or snakes; parasites, mosquitoes, or humans; or in the dividing cells of a tumor—the ongoing competition between predator and prey, pathogen and host, drug-resistant and susceptible cells, leads to changes in the genetic makeup of populations. This is the crux of evolution.

There can be no doubt about the power of natural selection to act on the smallest differences among individuals—small changes in the immortal genes have been eliminated in billions of species for 3 billion years, and a single change in a hemoglobin gene has enabled members of our species to survive malaria. But there is one aspect of the long-term action of natural selection I have not addressed—its cumulative creative power. Can natural selection on small variations really add up to forge the vast differences in complexity between life-forms?

Once again, the DNA record contains key new insights and the next chapter will complete the meal.

Many different kinds of animals build and inhabit the Great Barrier Reef of Australia. *Photograph by Antonia Valentine.*

Chapter 8

The Making and Evolution of Complexity

· · · · · · · · · · · · ·

The simplicity of nature is not to be measured by that of our conceptions. Infinitely varied in its effects, nature is simple only in its causes, and its economy consists in producing a great number of phenomena, often very complicated, by means of a small number of general laws.

—Pierre-Simon Laplace,
Exposition du système du monde (1796)

EVEN WITH A SNORKEL PROVIDING AMPLE AIR, THE sights below me take my breath away.

Drifting above the yellow, purple, and brown coral forest and boulders, the pageant of animal life surrounding the reef is a riot of colors, shapes, and sizes—schools of neon fish, brilliant sea stars, mottled octopi, spiny urchins, green sea turtles, black-tipped sharks, giant clams with vivid turquoise or magenta mantles, striped crabs, spotted rays, and cream-colored anemones.

The Great Barrier Reef of Australia is no doubt one of the great natural wonders of the world. Extending some

1200 miles along that continent's east coast, the Reef is the largest structure on Earth built by living organisms and the only one that is visible from the moon.

Such a great natural wonder has made many great naturalists wonder: How did the reef form? And how has such a great diversity of animal forms evolved?

In the early part of the nineteenth century, as geology found its footing and sought natural explanations for landforms, the prevailing view of coral reef formation was that they grew on top of submarine volcanic craters. The appearance of the idyllic atolls of the South Pacific, where circular reefs surround crystal blue lagoons, would seem to be accounted for by such a mechanism. But in geology and, as we will see shortly, in biology as well, looks can be deceiving. The submarine crater explanation of coral reef formation was overturned by . . . well, guess who?

If you said Darwin, well done.

Yes, twenty years before *On the Origin of Species*, in two books—first in his *Journal of Researches into the Geology and Natural History of the Various Countries Visited during the voyage round the world of H.M.S.* Beagle (1839) (commonly known as *The Voyage of the* Beagle) and then at length in *The Structure and Distribution of Coral Reefs* (1842)—Darwin provided a new explanation to account for the formation of all kinds of coral reefs, including the Great Barrier Reef. The significance of Darwin's coral reef theory is at least twofold. First, he was correct. Darwin's proposal was controversial and opposed for many decades, but eventually the evidence proved him right (again!). And second, his boldness and ability to theorize about a long-term process that no one had seen or could witness, and to extrapolate from small individual observations to comprehensive explanations, were a portent of the reasoning he would apply to the formation of the living world.

Darwin doubted the submarine crater theory because he thought it was unlikely that there were volcanic craters large enough to account for

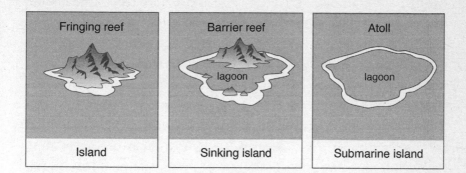

FIG. 8.1. **The formation of coral reefs.** Darwin proposed that the three major kinds of reefs form as successive stages of the same process. A fringing reef forms around a new landmass; as it subsides a barrier reef forms and a lagoon arises between the reef and land-mass. Finally, as the island sinks below the surface, an atoll forms with a reef surrounding a lagoon. *Drawing by Leanne Olds.*

certain large atolls, or that so many large volcanic craters would be crowded together beneath the sea where chains of atolls existed. Darwin also pointed out the crater theory was too atoll-centric and neglected two other common forms of reefs: the fringing reefs that surround oceanic islands, and barrier reefs that surround island lagoons. Instead, Darwin proposed a grand theory that fringing reefs, barrier reefs, and atolls represented three successive stages of the same process (figure 8.1). In Darwin's theory, a fringing reef first forms around the edges of a new island. Then, as the island sinks (subsides), the reef continues to grow, creating a barrier reef and lagoons around the island. Finally, as the island sinks below the surface of the ocean, an atoll forms.

The growth of most corals and the sinking of islands are impercep-tible—large boulder corals expand as little as 5 millimeters per year. But Darwin's appreciation for the cumulative effect of gradual change over vast periods of time, first derived from geology, gave him the con-fidence to theorize how large transformations occurred. Of course, twenty years after the coral reef theory, he provided a new explanation for the diversity of its many inhabitants.

Darwin's geological and biological theories involved extensive extrapolation—from small gradual change to large transformations, from the present to the past, and from simple forms to more complex ones. Much of the resistance to Darwin's theories was or is based on doubts about the validity of such extrapolations (e.g., not accepting the "adding up" of effects over vast periods of time). To this point in the book I, too, have implied a degree of extrapolation. I have shown how, for example, given an eye equipped with visual pigments, one or a few changes in pigment proteins alter their properties and help organisms adapt to different light environments. There should be no doubt about the reality of natural selection from the previous five chapters. That is a major purpose of this book—to eliminate any doubt. But, one may say, these are small changes in already complex structures—where did that eye come from in the first place?

It is a good and important question.

The evolution of complex structures has long been of central interest to biologists, and a refuge for evolution's doubters. It is not uncommon for individuals to acknowledge variation and evolution within species ("microevolution") but to refuse to extrapolate those processes to the formation of new species and to the evolution of complex features above the species level ("macroevolution"). Some states have gone so far as to paste stickers to this effect in students' biology textbooks (see page 240).

Darwin went to great lengths to explain how natural selection could shape organs "of extreme perfection and complication," such as the eye. Darwin's explanation was brilliant, but it was an extrapolation of the simple to complex, not founded on empirical knowledge of the history of the eye. The details of the building or evolution of such complex structures were far out of reach at the time, and for most of the next century.

This is no longer the case.

Over the past two decades, direct evidence has poured in as to how complex structures, particularly those of animals, are made and evolve. The catalysts of this new understanding have been advances in developmental biology, the science concerned with the processes through which a single-celled egg becomes a complex multi-billion- or trillion-celled animal. Development is intimately related to the evolution of form because all variations and changes in form arise through changes in development. The study of the evolution of development—dubbed Evo Devo for short—has led to many surprising and insightful discoveries about the evolution of complex bodies and body parts, in the process dismantling that refuge of evolution's doubters.[*]

In this chapter, I will highlight a few of the most important insights from Evo Devo into how the complex and diverse structures of animals have evolved. I will emphasize how understanding development reveals how complex structures are built and comparison of the development of the structures of different species reveals how complexity evolves. I will focus on a special group of body- and organ-building genes and part of the DNA record I have not yet introduced, but which is key to understanding the evolution of form.

Looks Are Deceiving: All Animals Share a Tool Kit of Organ- and Body-Building Genes

The animals I saw on the Great Barrier Reef included representatives of many major branches of the animal kingdom. Of the roughly thirty-five major animal groups, or phyla, there were cnidarians (corals, sea anemones), poriferans (sponges), mollusks (clams, octopi), arthropods (crabs), echinoderms (sea stars, sea urchins), and vertebrates (sharks, bony fish, sea turtles, and whales). Many of these animals are characterized by particular structures unique to their

---

*I wrote at length about many of these discoveries and their significance in my recent book *Endless Forms Most Beautiful: The New Science of Evo Devo*.

kind—the turtle's shell, the octopus's tentacles, the clam's shell, the crab's claws, etc.—but they also have organs that serve a similar purpose, such as eyes.

There is no doubt that eyes are useful to their owners. What has puzzled and intrigued biologists since Darwin's time is the variety of eye types in the animal kingdom. We and other vertebrates have camera-type eyes with a single lens. Crabs and other arthropods have compound eyes in which many independent unit eyes gather visual information. Even though they are not close relatives to us, octopi and squids also have camera-type eyes, while their clam and scallop relatives have among them three eye types: the camera-type eye with a single lens, a mirror eye with a lens and a reflecting mirror, and compound eyes made up of from ten to eighty unit eyes.

The great diversity of eye structures was, for more than a century, believed to be the product of many independent inventions of eyes in different animal groups. The great evolutionary biologist Ernst Mayr and his colleague L. V. Salvini-Plawen suggested, on the basis of the cellular anatomy of animal eyes, that eyes had been invented independently some forty to sixty-five times.

On the one hand, this would seem to support the notion that evolution can repeatedly come up with solutions to the same need (vision). This view of the repeated evolution of eyes was widely accepted. But new discoveries have forced a reexamination of eye evolution. The key issue centers on whether eyes were repeatedly invented *from scratch*, or whether the evolution of eyes has involved the use of common ingredients present in one or more common ancestors. These two possibilities weigh on our notions of the plausibility and probability of the evolution of complex structures. Obviously, it would appear to be more "difficult" (less likely or frequent) to evolve a structure from scratch than from parts that are already available. The new evidence reveals that vastly different eyes have much more in common than anyone realized, and these common ingredients give us powerful insights into how complex structures evolve.

Pax-6 protein sequences

Fruit fly LQRNRTSFT<u>NDQIDS</u>LEKEFERTHYPDVFARERLA<u>GKIG</u>LPEARIQVWFSNRRAKWRREE

Mouse   LQRNRTSFTQEQIEALEKEFERTHYPDVFARERLAAKIDLPEARIQVWFSNRRAKWRREE

Human   LQRNRTSFTQEQIEALEKEFERTHYPDVFARERLAAKIDLPEARIQVWFSNRRAKWRREE

FIG. 8.2. **The *Pax-6* eye-building gene.** A portion of the amino acid sequence of the fruit fly, mouse, and human proteins is shown. Note the great similarity between the fly and mammal proteins, and that mouse and human sequences are identical.

The new story of eye evolution began to unfold in 1994. Walter Gehring and his colleagues at the University of Basel (Switzerland) were studying a gene required for the formation of the compound eyes of fruit flies. When this gene was inactivated by mutations, eye formation was abolished; earlier fly geneticists had dubbed the gene *eyeless* (many genes are named for what goes wrong when they are mutated; this gene's normal job is to promote eye formation). When the DNA encoding the *eyeless* gene was isolated, the researchers discovered, much to everyone's surprise, that the *eyeless* gene encoded a protein that was remarkably similar to proteins encoded by two genes that had been discovered in mice and humans. The mouse protein was called Small eye and it, too, was required for the formation of eyes— the camera-type eye of mice. The human protein was known as Aniridia, because defects in it caused a condition in which the iris or larger parts of the eye were missing. The similarities between the fly, mouse, and human proteins are so strong that we know we are looking at the same protein in these different species (figure 8.2). The protein is now referred to by a single, less descriptive name, Pax-6.

The immediate question raised by the discovery of Pax-6's involvement in the formation of such different eyes as those of the fly and of mammals is whether this is mere coincidence or a sign of something of greater significance. In other words, did Pax-6 just happen to get used by flies and mammals in the evolution of their eyes from scratch, or are different appearing eyes more closely related than we suspected and their construction under the command of Pax-6 is a sign of some deep common principle?

Many more discoveries have now weighed in on these alternatives. The first were experiments that showed that the mouse and fly *Pax-6* genes were interchangeable in fly eye development. Using special techniques, the Swiss researchers activated the fly *Pax-6* gene in unusual places like the legs, wings, or antennae and showed that it could induce the formation of eye tissue! They then discovered that the mouse *Pax-6* gene could also induce fly eye tissue. So clearly, the genes had the same capabilities, not just similar sequences. Remember from chapter 3 that nothing encoded by DNA can endure over time without natural selection. For some reason the function and sequence of the Pax-6 protein have been preserved across a vast span of time of animal evolution—over 500 million years.

The reason for the *Pax-6* gene's preservation is made clear by a second set of observations concerning its use for building eyes in other animals. *Pax-6* has also been isolated from a squid, a planarian, and a ribbon worm and shown to be used in the development of each creature's complex or simple eyes as well.

Since Pax-6 is involved in eye development in such a wide range of animals, it is very unlikely that each happened upon the use of Pax-6 by accident. The widespread role of Pax-6 in eye development must be due to historical reasons. That is, a common ancestor of these animals used Pax-6 in the development of some, perhaps very primitive eye. All the marvelous and more complex eyes that evolved in this ancestor's descendants would then be built on this foundation.

The next interesting question then, which is central to our picture of how complexity evolves, is: What was this foundation? What ingredients were already in place in animal ancestors and available for evolving more complex eyes?

We know much more about the ingredients of eye development and function than just the role of Pax-6. All eyes are composed of the light-detecting cells known as photoreceptors and of pigment cells that govern the angles of light that reach the photoreceptor cells. Thus, it is most reasonable to deduce that the most primitive eye was composed of these two cell types. This is just what Darwin inferred: "The simplest organ which can be called an eye consists of an optic nerve, surrounded by pigment-cells and covered by translucent skin, but without any lens or refractive body."

Such a simple, two-celled eye is, in fact, found in the larvae of creatures such as the marine ragworm, *Platynereis dumerilii*. After its first day of development from a fertilized egg, the larvae has a pair of two-celled eyes "gazing out" from its front end (figure 8.3, top row). But do not be fooled by these eyes' simple construction and appearance. They are built with and use many of the ingredients used in fancier eyes. For example, the detection of light in the photoreceptor cells of this simple eye relies on opsin proteins, the same visual pigments discussed in several previous chapters. Indeed, all animal eyes utilize opsins for light detection. The inescapable explanation is that opsin existed in a primitive eye in the common ancestor of most animals, and has since been used for detecting light in all varieties of eyes.

A good picture of how more complex eyes are built and evolve also emerges from these ragworm larvae. The development of the larger, cup-shaped adult eye begins near the simple two-celled larval eyes and is assembled with many more photoreceptors and pigment cells (figure 8.3, bottom row). Complexity, in this case, is a matter of just arranging larger numbers of the same types of eye cells in three-dimensional space—the same building materials, a different organization. And the same tools are used. Pax-6 and at least two additional types of eye-building genes known from flies and vertebrates are involved in this process in the ragworm. The construction of larger, but still primitive, adult eyes from these basic eye cell types in the ragworm, and the use of the same genes as those used in building more elaborate compound and camera eyes gives us a picture of how com-

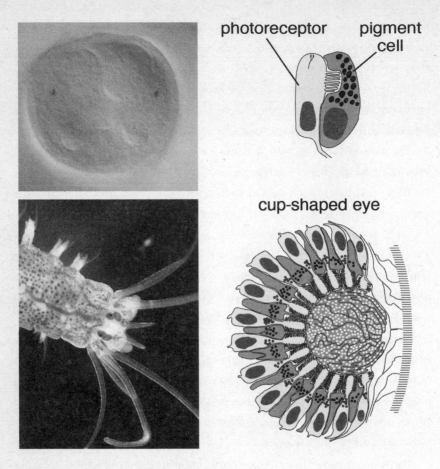

FIG. 8.3. **Simple and more complex eyes of a marine worm.** A simple pair of eyes forms in the day-old larvae of the ragworm (top left), each of which is composed of just two cells (top right). In the adult worm, two pairs of eyes form (bottom left) that are composed of many more cells arranged in a cup shape (bottom right). Some of the same genes build each kind of eye. *All panels courtesy of Detlev Arendt, European Molecular Biology Laboratory, Heidelberg, Germany (adapted from Arendt et al. [2002], Development 129:1143) used by permission of the Company of Biologists Ltd., except for bottom left, which is courtesy of Benjamin Prud'homme, Howard Hughes Medical Institute and University of Wisconsin.*

plexity is built and evolves. We can see, in development, that more complex organs are constructed by assembling larger numbers of a few cell types, and that in evolution, the same cell types and eye-building genes have been used in constructing modern eyes. In different animals, the same cellular building blocks and genetic "tools" have been used to build eyes with different structures.

The new picture of eye evolution reveals different kinds of eyes as products of different evolutionary journeys that began at similar starting points, with some simple arrangements of photoreceptors and pigment cells, and not from scratch. Nor is it the case that camera eyes evolved from compound eyes or vice versa. If one thinks only about modern, complex forms, one might try to derive one type of eye from another. It is hard to see how that conversion could occur without a decline in performance in the intermediate stages. But that is not what transpired in evolution.

Instead, the history of eye evolution now appears to be one of the repeated evolution of more complex eyes from simpler proto-eyes (figure 8.4). The role of natural selection in the evolution of complex eyes with better optical qualities is easy to explain. From simple beginnings, incremental changes that improve eyes' functional properties would be incorporated. Indeed, if we look at just one phylum of animals, the mollusks, we find a wide variety of eyes that represent different grades of complexity (figure 8.5). Computer modeling by Dan Nilsson and Susanne Pelger at the University of Lund in Sweden has suggested that selection on small variations could, in 2000 steps over as few as 500,000 years, produce a camera eye from a simple prototype.

The expanding picture of eye evolution has also helped to explain some of the peculiar and interesting differences among eye types. For example, in our eyes the photoreceptors point away from the light and are at the back of the eye, while in squid eyes they point toward the light and are at the front of the eye (see figure 8.4). It is extremely difficult (and unnecessary) to imagine how one arrangement could

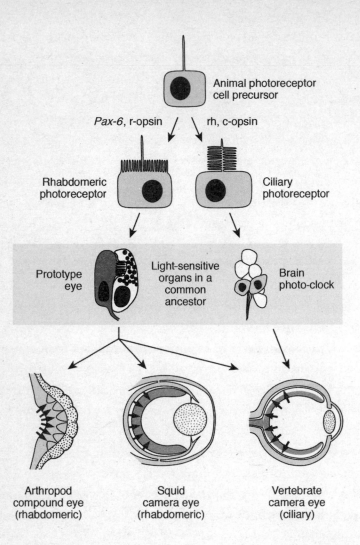

FIG. 8.4. **The origin and evolution of complex eyes.** Photoreceptor cells evolved in animal ancestors under the control of the *Pax-6* gene and detect light through opsins. Complex eyes evolved from simpler arrangements of photoreceptor and pigment cells. A common ancestor of bilateral animals possessed two kinds of photoreceptor cells, a rhabdomeric type involved in vision in a prototypic eye, and a ciliary type involved in a light-sensitive clock in the brain. Rhabdomeric photoreceptors were recruited into the evolution of arthropod and cephalopod eyes, and both photoreceptor types were recruited into the evolution of vertebrate eyes. *Drawing by Leanne Olds.*

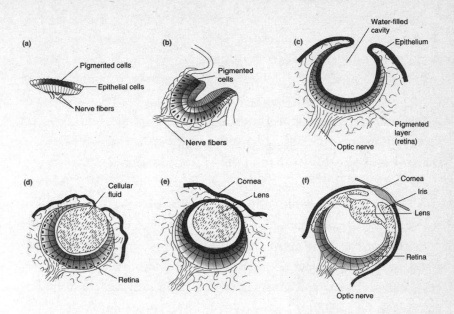

FIG. 8.5. **Various stages of eye evolution found in mollusks.** The architecture of mollusk eyes ranges from a simple pigmented eye-spot (a), to a cup-shaped eye (c), a cellular fluid-filled eye (d), a lens-covered eye (e), and the complex eye of squids (f). *Figure from M. W. Strickberger,* Evolution, *copyright ©1990 by Jones and Bartlett Publishers, Boston; used by permission.*

be derived from the other. Clearly, there is more that one way to evolve a camera-type eye and cephalopods and vertebrates hit upon different solutions.

One of the other deep puzzles of eye evolution concerns the different kinds of photoreceptors found in our eyes versus those found in squid or fly eyes. In human and other vertebrate eyes, the rod and cone photoreceptor cells are of the so-called ciliary type, while in squids and flies, the photoreceptor cells are called rhabdomeric. The distinction has to do with how the membranes of each type of photoreceptor cell are enlarged in order to pack them with opsins. This difference was one of the key pieces of cellular evidence cited for the "independent" origin of vertebrate and other animal eyes.

New discoveries, again from the humble ragworm, have clarified the mystery of the origin of our eyes and photoreceptors. Detlev Arendt and colleagues at the European Molecular Biology Laboratory (EMBL) in Heidelberg, Germany, noticed some ciliated cells in the developing brain of the ragworm that bore an uncanny resemblance to vertebrate photoreceptors. Further study revealed that these cells expressed a particular opsin that bore greater similarity to the opsins of vertebrates than to the opsins found in the photoreceptor cells of the ragworm's eyes or other invertebrate eyes. The "ciliary" brain opsin (c-opsin) appears to be involved in the control of the worm's daily biological clock, not vision. The EMBL researchers demonstrated that the ragworm, an invertebrate annelid worm, possesses both known types of photoreceptors and opsins. This discovery suggests that a common ancestor of the ragworm, squid, and vertebrates also had both kinds of photoreceptors and opsins. The rhabdomeric type and its opsin (r-opsin) were recruited into the visual system of arthropods and cephalopods while the ciliary type and its opsin were recruited for vertebrate vision. (It also appears that vertebrate eyes recruited the rhabdomeric cells into becoming so-called retinal ganglion cells, which function in the output of the retina; thus the vertebrate eye appears to be derived from both photoreceptor cell types.)

The eye, far from being one of the most difficult structures to account for by evolution, has become instead one of the leading sources of insights into how evolution works with common genetic tools to build complex organs. Discoveries in Evo Devo have also revealed that common genetic tools are used to build the very different hearts, digestive tracts, muscles, nervous systems, and limbs of all sorts of animals. It is clear that just as photoreceptors are an ancient cell type, the types of cells that make up many tissues and organs are ancient. Furthermore, we now know from gene and genome sequencing that most animals are endowed with similar tool kits of body-building and organ-building genes (our phylum, the vertebrates, does have a larger number of these tool-kit genes because of some large-scale genome duplications). This tells us that the tool kit itself is ancient and must have been in place in a common ancestor before most types of modern animals bodies and body parts evolved.

The identity of that common ancestor is not known. But if asked to sketch what that creature might look like, a small, soft-bodied, free-swimming marine animal like the ragworm larva (figure 8.3, top left), with its full genetic tool kit, many cell types, and simple organs, would be a good approximation of the foundation upon which most of the animal kingdom was built.

These new discoveries about development and tool-kit genes help us to see directly into the process of how complex structures are made and evolve. But they also present a paradox, namely: With so many cell types and body-building genes in common, how do such vast differences in form evolve?

Diversity Is the Product of Using Similar Body-Building Genes Different Ways

Before I delve further into the evolution of form, it is important to underscore a critical distinction between Pax-6 and other tool-kit proteins, and the sorts of proteins I have discussed in previous chapters. Opsins, globins, ribonucleases, odorant receptors, etc., are proteins that are directly responsible for *physiology*—of vision, respiration, digestion, and olfaction. Pax-6 and other tool-kit proteins are devoted to the making of *form*—they control the number, size, and shape of body parts, as well as the identity of cell types in the body. Most tool-kit proteins act, directly or indirectly, by controlling when and where many different genes are used in the body. The dramatic effects of Pax-6—the loss of eyes when it is inactivated, the induction of eyes where it is active—are due to its effects on many other genes and steps in development. Furthermore, Pax-6 and most other tool-kit proteins have many jobs in building the body and body parts. Pax-6, for example, is also involved in building part of the brain and nose in mammals. Some tool-kit proteins may work in building or shaping ten, twenty, or even more body parts.

One absolutely crucial difference, then, between proteins involved in physiology and those involved in body-building, concerns the consequences of mutations that alter these proteins. A mutation that alters an opsin protein may affect the spectrum of light detected in either rods or cones in the eye. However, a mutation in a tool-kit protein may abolish the eye altogether, as well as affect other body parts. For this reason, mutations that alter tool-kit proteins are often catastrophic and have no chance of being passed on. The important consequence is that the evolution of form occurs more often by changing how tool-kit proteins are used, rather than by changing the tool-kit proteins themselves.

I will describe two examples that demonstrate how the evolution of form often occurs through changes in parts of DNA that do not encode proteins, but contain the instructions that govern how tool-kit genes are used. This less explored and much less well understood part of the DNA record holds the keys to understanding how such great architectural diversity has been generated using a common set of tools.

One of the most obvious features of large, complex animals is their construction from repeated parts. Just as tissues and organs are made up of cellular building blocks, animal bodies are often constructed from building blocks. For example, segments are the building blocks of arthropods (insects, spiders, crustaceans, centipedes), and vertebrae are the building blocks of our backbones and those of other vertebrates. Many structures are reiterated that emerge from these blocks, such as the many appendages of arthropods (legs, claws, wings, antennae, etc.) and the ribs of vertebrates. One of the widespread trends in the large-scale evolution of these animals' bodies concerns changes in the number and kind of repeating parts. The major features that distinguish classes of arthropods are the number of segments and the number of kind of appendages. Similarly, classes of vertebrates differ in the number and kind (cervical, thoracic, lumbar, sacral) of vertebrae.

Differences in the number and form of repeating parts are not restricted to higher taxonomic ranks, but can be found among closely related species or populations. For example, the threespine stickleback fish occurs in two forms in many lakes in North America—a shallow-water, bottom-dwelling, reduced-spined form and an open-water, full-spined form (figure 8.6). The pelvic spines are actually part of the fishes' pelvic fin skeleton, and the pelvic and pectoral fins are repeated structures. Pelvic spine length is under selection pressure from predators. In the open water, longer spines help protect the fish from being swallowed by larger predators. But, on the lake bottom, long pelvic spines are a liability. Dragonfly larvae seize and feed on young sticklebacks by grabbing them by their spines.

The evolution of these stickleback populations is very recent. The lakes they inhabit were formed by receding glaciers in the last ice age, approximately 10,000 years ago and each lake was colonized by oceanic sticklebacks that then rapidly and repeatedly diverged into the short- and long-spined populations. Exceptional fossil records of stickleback evolution have been uncovered that document their rapid evolution.

Because the two populations are so recently evolved, they can still mate together and produce offspring. This allows geneticists to trace the genetic changes that underlie the divergence of body forms. Recently, David Kingsley, Dolph Schluter, and their collaborators at Stanford University and the University of British Columbia have been able to pinpoint genes responsible for the evolution of different traits in sticklebacks. The evolution of one trait, the pelvic spines, reveals how the formation of a repeated structure evolves through changes in the way a tool-kit gene is used.

The reduction of pelvic spines in bottom-dwelling populations is due to a reduction in the development of the pelvic fin bud. The major gene responsible for the reduction of the pelvic skeleton was recently identified as a tool-kit gene called *Pitx1*. This is a typical tool-kit gene—it has several jobs in the development of the fish, it acts by controlling other genes, and has counterparts in other animals, such as the mouse. In the mouse, *Pitx1* helps make the hindlimb different from the forelimb (limbs are another repeated structure).

We know from the fossil record that the pelvic fin was the evolutionary forerunner of the hindlimb of four-legged animals. The use of *Pitx1* in the development of the pelvic fins in fish and in mammal hindlimbs is a very nice, independent piece of evidence supporting that history. But the main point I want to make here is how the fishes' pelvic skeleton gets reduced by changes at the *Pitx1* gene without affecting other body parts where *Pitx1* also functions.

The big clue comes from comparing the Pitx1 proteins of the pelvic-reduced and full pelvic forms. There is not a single difference in the protein sequence.

But, wait, didn't I say that changes at *Pitx1* made the pelvic skeleton different? Yes, I did. The apparent paradox is resolved by understanding that, in addition to the coding part of a gene, every gene also contains noncoding DNA sequences that are *regulatory*. Embedded in this regulatory DNA are switchlike devices that determine where and when each gene is or is not used. Tool-kit genes can have many separate switches, with each switch controlling the way a gene is used in a different body part. The function of switches depends on their DNA sequence, and changes in their sequence can alter how they work. A critical property of these switches is that changes in one switch will not affect the function of the other switches. And therein lies a huge insight into how form evolves. That is, the use of a tool-kit gene can be fine-tuned in one structure without affecting any other structures.

In the pelvic-reduced stickleback, the *Pitx1* gene is, in fact, not used in pelvic fin development. Changes in the switch that govern its use in the hindlimb have enabled the selective reduction of this part of the

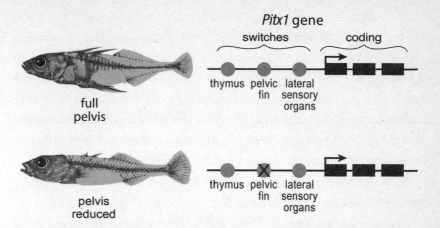

FIG. 8.6. **The evolution of the stickleback fish pelvic skeleton.** Two forms of stickleback fish occur in many lakes, the bottom-dwelling form has a reduced pelvic skeleton. The reduction of the skeleton is due to a change in the function of a genetic switch controlling the use of the Pitx1 gene in the developing pelvic fin (X). *Drawing by Leanne Olds.*

fishes' skeleton (figure 8.6). The power of this example lies in its demonstration of how, at the fundamental level of DNA, a major change in body anatomy can rapidly evolve.

The reduction of the hindlimb has happened several times in vertebrate evolution. Cetaceans and manatees have evolved greatly reduced hindlimbs as they evolved from land-dwelling ancestors into aquatic forms, and snakes and legless lizards have also reduced their limbs. The sticklebacks, and a large body of work on other animals that I will not describe here, are revealing how such changes in body form and complex structures are achieved.

The reduction and loss of structures is just one facet of the evolution of form. Of course, we also want to know how new features evolve, and the evolution of switches is again center stage.

Endless Flies Most Beautiful

While tropical fish, butterflies, and birds probably hold the top rankings for color and beauty in the animal kingdom, in the kingdom of biological research few animals rival the fruit fly. The discovery of fruit fly body-building genes sparked the resurgence of developmental biology and the emergence of Evo Devo. More recently, the diversity of fruit fly wing patterns, albeit less colorful than those of birds or butterflies, has helped illuminate how new traits evolve.

The laboratory fruit fly *Drosophila melanogaster* has pale wings but its many cousins in the large family to which it belongs display a great variety of black pigment patterns (figure 8.7). In many species, the pigment patterns are restricted to males and are displayed during elaborate courtship dances as the male sways or preens in front of the female. A common pattern is a black spot near the tip of the wing.

In my laboratory at the University of Wisconsin, we have been tracing how such spots are made and evolve. These spots are an excellent illustration of the general phenomenon that new patterns evolve when "old" genes learn new tricks.

The making of the black spots involves enzymes that synthesize melanin, the black pigment described in chapter 7. Think of these enzymes as paintbrushes in the tool kit. In species that have black patterns on the wing, these paintbrushes are used in patterns that foreshadow the black pigment areas. The patterns are controlled by switches that surround the coding part of each paintbrush gene. In the evolution of black spots on the wing, several changes have occurred in switches of paintbrush genes. The evolution of a dark black spot with sharp borders was not a one-step "now you don't see it, now you do" event, but a multistep series of changes during which the intensity and the shape of the spot evolved. Thus, this "simple" spot is really a complex character, built up over time by the "adding up" of many variations—as we think most physical traits are.

FIG. 8.7. **The great diversity of fruit fly wing patterns.** These little insect wings provide a great example of how seemingly endless varieties of patterns evolve using the same sets of genetic tools. *Montage by Nicolas Gompel and Benjamin Prud'homme.*

We have pinpointed changes that have occurred in a switch that governs how one paintbrush gene is used in the wing (figure 8.8). The paintbrush gene has other separate switches that control how it is used in other body parts (such as the thorax and abdomen) or at other stages of development (such as in the fly larva). So, again, the existence of separate switches allows the use of a tool-kit gene to be modified in one part of the body without affecting how the gene is used in other body parts. In other species, different switches have been modified in making other patterns.

Once this black spot evolved, it was inherited by many descendant species. However, curiously, it has also been lost by a handful of species. The loss of traits is a more common trend in evolution than many people realize. For the wing spot, it may be that the females stop selecting on the trait and this relieves the pressure to maintain it, so it disappears from males. We have investigated the mechanism of spot disappearance and discovered that the switch that made the spot in the ancestor has accumulated mutations that inactivate it—much like the fossil genes I described in chapter 5. Just as proteins can be lost by mutations, switches can also be inactivated. The neat difference when a switch is inactivated is that the paintbrush gene continues to work to color other body parts.

By tinkering with the switches of pigmentation genes, flies have evolved a great diversity of wing patterns under both sexual and natural selection. The importance of these little insect wings is that they illustrate how a seemingly endless variety of patterns can be generated using the same tool kit of body-building and body-painting genes, by tinkering with genetic switches.

The Making of Complexity

I opened chapter 1 and the main body of this book by quoting Darwin's conviction that "when we regard every production of nature as one which had a history; when we contemplate every complex

Paintbrush gene

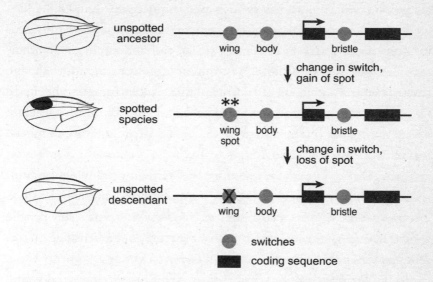

Fig. 8.8. **The gain and loss of wing spots occurs through specific switches in a paintbrush gene.** The evolution of a spotted wing from an unspotted ancestor involved the modification of a genetic switch controlling the use of a black "paintbrush" gene in the developing wing (asterisks). The selective loss of the wing spot (without any effect on other body parts) occurred by mutations in the wing spot genetic switch (X). *Drawing by Leanne Olds.*

structure and instinct as the summing up of many contrivances, each useful to the possessor . . . how far more interesting, I speak from experience, will the study of natural history become!"

Darwin's use of the now peculiar-sounding term "contrivance" was a deliberate choice for rhetorical effect. Darwin was evoking the same terminology used by Reverend William Paley in his well-known book *Natural Theology* (1802). Paley saw the fashioning of contrivances in nature for specific purposes as revelations of God's design: "Contrivance must have had a contriver, design, a designer." Darwin, who actually admired Paley's book and declared he had committed most of it to memory, fashioned much of his argument in *On the Origin of Species* as a direct refutation of Paley's argument for design.

Darwin's great advantage over Paley and other thinkers of his generation was his grasp of the immensity of time. His aptitude for geology gave him a sense of both the cumulative power of gradual change and the enormous spans of time available. His imagination or, more important, his grasp far surpassed those straitjacketed by biblically based interpretations of the age of the Earth. In order to overcome the entrenched resistance and doubts of his audience and the population at large, Darwin knew that he must assemble a prodigious body of facts, find the most apt analogies and metaphors, and use the most persuasive prose. He anticipated the particular difficulties people would have in picturing how complex structures or contrivances arose. But he also knew the deep rewards in doing so.

Today, the body of facts concerning evolution continues to expand in all dimensions. The great advances in developmental biology, Evo Devo, and the deciphering of the DNA record provide a whole new vista into both process and history. Developmental biology offers a window on the making of complexity on an easily observable timescale. The whales, turtles, fish, crabs, and corals of the barrier reef are complex, but they all start life as a fertilized egg and in a matter of just days, weeks, or months, a complete individual, with its many complex parts, is built by processes that we are now understanding in great detail. Evo Devo links the differences among species in the everyday process of development to the long-term process of the evolutionary change of form—the "summing up" of all of the changes accumulated in development over many thousands or millions of generations. And the DNA record enables us to reconstruct individual steps in evolution.

The argument for design by some external intelligence is eviscerated.

It is hard to imagine how anyone in command of these facts could harbor any reasonable doubt. These facts are derived from the very same science and technology that has deciphered the genetic cause of hundreds of diseases, invented dozens of new gene-derived medicines, and revolutionized forensics and agriculture. Yet, in the face of all of the evidence, there remains much doubt and outright denial of the reality of biological evolution. To understand this doubt and denial, we'll leave the realm of scientific evidence, because the reasons for such doubt could not be, and are not, scientific. They are cultural. And they can best be understood in light of the many instances in history when certain groups, out of self-interest and ideology, have denied new scientific knowledge.

It is time to move on to the afterdinner conversation.

Statue of Louis Pasteur at Place de Breteuil, Paris, France.
Photograph courtesy of Benjamin Prud'homme.

Chapter 9

Seeing
and Believing

· · · · · · · · · · · · ·

Thinking is seeing . . . every human science is based on
deduction, which is a slow process of seeing by which we
work up from the effect to the cause.
—Honoré de Balzac, *La Comédie humaine* (1845)

IT WAS CALLED THE HOUSE OF CRIME.

The clinic in the French countryside lost so many of its
female patients to infection, sixteen out of sixteen in one
stretch, that its founder declared that whoever solved the
problem deserved to be honored with a gold statue. At the
Academy of Medicine in Paris, the profession was stumped.
As one physician scoffed at the notion of disease being
spread by doctor's hands, an audience member rose in anger,
"The thing that kills women . . . is you doctors that carry
deadly microbes from sick women to healthy ones."

The man in audience was Louis Pasteur.

This was 1879, thirty years after the physicians Oliver
Wendell Holmes and Ignaz Semmelweis had suggested
handwashing to prevent "childbed fever." It was more than
twenty years after Pasteur had demonstrated that the air

was full of microbes that would grow under the right conditions, thus debunking the widely accepted myth of spontaneous generation. The germ theory had still not yet been accepted nor had its simplest application been adopted.

Some European doctors and clergy thought that childbed fever was God's punishment to women for the act of childbirth. Changing this point of view meant accepting not only that the lethal infection was caused by something invisible, but that the doctors themselves were carrying it to their patients.

This stubborn disbelief could be overcome only with more evidence, and that was what Pasteur and several contemporaries provided. The German Robert Koch used new, more powerful microscopes to see and identify the microbes that caused anthrax, cholera, and tuberculosis. Joseph Lister, a Scottish surgeon, extrapolated from Pasteur's work on germ theory and fermentation to invent means of sterilizing wounds and instruments, and thus reduced the mortality from surgery by over 70 percent.

Pasteur's own demonstrations were perhaps the most persuasive. In the 1870s as many as half of all sheep and cattle in France were dying of anthrax. Despite great difficulties in culturing the bacterium, Pasteur succeeded in producing a vaccine and demonstrated its success in the field in 1881.

The germ theory was the foundation of many improvements to human health, including pasteurization, as well as the prevention and control of infectious diseases. Pasteur is honored not just for these contributions but for his embodiment of the scientific method. He had a great ability to survey all of the available data and to incorporate it into hypotheses, he was driven to design rigorous experiments that would test them, and he had the creative spark to synthesize theory and experiment into new knowledge.

Pasteur articulated the critical role of experimental evidence:

Imagination should give wings to our thoughts but we should always need decisive experimental proof, and when the moment comes to draw conclusions and to interpret the gathered observations, imagination must be checked and documented by the factual results of the experiment.

In hindsight, of course, it might seem incredible that the medical profession sustained their ignorance for so long. But keep in mind that, for much of this time, the doctors were being asked to believe something they could not see. For them, and most of the rest of us, seeing *is* believing. We believe what we witness with our own eyes. Throughout scientific history, new ways of seeing have played a critical role in the discovery and the acceptance of new ideas. The proof of the germ theory—in the microscope, in the farmers' fields, and in the clinic—removed all doubt.

Charles Darwin certainly subscribed to Pasteur's doctrine of scientific research but Darwin had the disadvantage that his great theory would not be so readily tested nor so convincingly proven during his lifetime. Like Pasteur, Darwin postulated about invisible forces in nature, but he couldn't inject something into an animal and make it evolve into another form, nor look in a microscope and see evolution happen. Darwin drew upon all of the evidence he could muster, from geology, fossils, the breeding of domesticated animals, and a wealth of naturalist studies of plants and animals to synthesize and support his theory. However, the timescale of evolution put the direct observation of a species changing out of reach. And, as with the germ theory, there was stubborn disbelief.

Now, nearly 150 years later, we can see. We no longer look at Nature's diversity "as a savage looks at a ship." From the new DNA record, the evidence of the workings of the evolutionary process abounds. But many people—a great many—either do not see what scientists see, or do not believe what scientists have concluded.

I have borrowed the title of this chapter from the wonderful book *Seeing and Believing* by Richard Panek, about the invention of the telescope and how it changed our perception of the sky and our place in the universe. Like Darwin, Galileo's observations and ideas were rejected by authorities who had no use for new evidence or ideas. But, eventually, the observable evidence overwhelmed ideological resistance. For all of those who do see the overwhelming evidence of natural selection and life's descent from ancestors, and the immense span of time over which the story of life has unfolded, it is, to put it mildly, baffling how so many still do not. It is absolutely astonishing and often infuriating that some take it so far as to deny the immense foundation of evidence and to slander all the human achievement that foundation represents.

With the facts on evolution's side, how can this doubt and denial persist, or even be growing, here at the outset of the twenty-first century?

I will confront the denial of evolution later in the chapter. Before I do, I want to equip you with some fresh understanding of the motivation for and tactics of denial, by examining other instances in history where scientific facts and evidence were actively ignored. The Galileo episode is usually trotted out as exhibit A in the history of conflict between science and religious orthodoxy. That story has been repeated often enough and that period is probably too far removed from the present for most of us to relate to it. Furthermore, while the resistance to evolution is religiously motivated, this is not always the basis for resistance to scientific knowledge. Nor are all religious denominations at odds with evolutionary science—in fact, far from it. Instead of rehashing that ancient history, I will describe two fascinating and illuminating, but not very widely known, recent examples of holdouts against twentieth-century biology. The first story involves the denial of the science of genetics, and later of DNA as the basis of heredity, by biologists who held power in the Soviet Union from the 1930s to the 1950s and beyond. These zealots essentially destroyed biology in the USSR, with tragic consequences for science, scientists, and the populace. The second involves the chiropractic profession and its denials, initially of the germ theory of disease, and then its subsequent and still active opposition to the science and medical value of vaccination.

The French doctors, Soviet biologists, and chiropractors I indict here placed ideology and self-interest above the pursuit and applica-

SEEING AND BELIEVING   *219*

tion of true scientific knowledge, as do the denyers of evolution today. Louis Pasteur once said, "Knowledge is the heritage of mankind." Protecting and advancing that heritage requires that we understand the motivations and strategies of those who would threaten it, and when needed, expose and vigorously oppose their misguided agendas.

The Commissar of Soviet Biology

Trofim Denisovich Lysenko (1898–1976), a peasant with a meager education, rose to become a deputy of the USSR Supreme Soviet, a full member of three scientific academies, and director of the Genetics Institute of the USSR Academy of Sciences. He received the Stalin Prize three times, was named a Hero of Socialist Labor, and was awarded the Order of Lenin, the country's highest honor, *eight* times. For more than twenty-five years T. D. Lysenko reigned over Soviet biology, agriculture, and medicine.

And he destroyed it.

The story of T. D. Lysenko has many chapters. Several books have been written that chronicle his ascent to prominence and power, his political intrigues and contribution to the horrors of Stalin's purges, and his long period of influence over Soviet biology, despite his profound ignorance and shoddy scientific skills. Two of the most moving accounts were written by scientists who were victims of Lysenko's influence or of the Soviet political system. Zhores Medvedev's *The Rise and Fall of T. D. Lysenko* and Valery Soyfer's *Lysenko and the Tragedy of Soviet Science* are sobering works written by very courageous men who made enormous personal sacrifices for the cause of truth.

I cannot do justice to this epic tragedy, or to the suffering of the many scientists who resisted Lysenko, in my brief account here. Instead, I will focus on a few key turning points when good science was ignored for ideological or political reasons, and I will describe how political motives eviscerated the body of Soviet biology. From a Westerner's view, the scope of denial in the Lysenko era of Soviet biology is incomprehensi-

FIG. 9.1. **Trofim Denisovich Lysenko**.

ble, and the consequences of not supporting that denial are unimagin-
ably dire. But the cultural influences and individual concerns that fos-
tered this denial are not at all unique or restricted to the past.

Genetics as a Bourgeois Science

In the early part of the twentieth century, after the rediscovery of
Mendel's work, fundamental work on the nature of genes and inheritance was led by T. H. Morgan. For his seminal discoveries, made
studying fruit flies, Morgan was awarded the Nobel Prize in 1933. The
understanding of genes as particulate units within cells, subject to
mutation, explained both the constancy and variation of species.
Genetics was recognized as a key science all over the world, including
the Soviet Union.

And then came T. D. Lysenko.

Lysenko's first brush with fame came from a rather simple assignment he was given while working as an assistant at the Gandzha Plant Breeding Station in Azerbaijan. The director was Nikolai Vavilov, one of the best known and most accomplished biologists in the USSR at the time. Vavilov had studied under William Bateson, the British scientist who brought Mendel's work to widespread attention and who coined the term "genetics." Vavilov traveled throughout the world, amassed a unique botanical collection, and enjoyed an international reputation. One of the major imperatives for agriculturists in the post-revolutionary USSR was the acute need to improve crop yields. The Soviet Union suffered perilous declines in grain harvests and in livestock numbers during the collectivization of agriculture (1928–1932).

Lysenko tested the idea that plants from more northern latitudes, in this case peas, might survive the winter and provide fodder for spring plantings of cotton. Fortunately for Lysenko, the first year brought a mild winter and promising success. A journalist for *Pravda*, the Soviet Communist Party newspaper, sensationalized the results and singled out Lysenko as a "barefoot professor" who "now has followers, pupils, an experimental field; and the luminaries of agronomy visit in the winter, stand before the station's green fields, and gratefully shake his hand." This simple peasant learned a valuable lesson in propaganda. The experiment, however, was not successfully repeated and future winter pea crops failed.

Lysenko then turned his attention to "vernalization," a method for obtaining winter crops by soaking the seeds from summer plantings at low temperatures. He received more attention from *Pravda* when his father planted vernalized wheat on a plot in his village. Great claims were made as to the success of the "experiment," heralding it as an "extraordinary discovery . . . corroborated by outstanding experimental data" whose "prospects . . . are so great they are beyond immediate calculation." Valery Soyfer concludes from the available evidence that the reports of a large harvest were false and explains the impetus to accept such claims:

It was as though all at once they believed in the power of a miracle and decided that they had the firebird in their hands. Indeed it was no accident that they were seduced by myths promising to solve all problems at one stroke, rather than facing the real and serious work of putting agriculture on a sound foundation. All Soviet life was permeated by myths, by expectations of the imminent approach of communism, a bright future just ahead. The notion of ordinary working people achieving miracles . . . perfectly matched the mood.

Lysenko promoted vernalization, and was in turn promoted to a newly created department for vernalization at an institute in Odessa. There he began to develop a theory to explain vernalization. Lysenko postulated that hereditary changes occurred in plants in response to environmental influences. This idea was a fully Lamarckian theory, the notion that organisms could pass on to their offspring characteristics acquired during their life. This theory had great philosophical appeal in the political climate of the Soviet struggle, for it suggested that Nature, and man, could be shaped in any desirable way, and were not constrained by history or heredity.

But Lysenko's ideas flew in the face of emerging understanding of genetics. And so, the die was cast. The conflict between Lysenko and geneticists, including his former mentor Vavilov, would define the next two decades of Soviet biology.

Lysenko's fame spread. The commissars of agriculture and Party functionaries heard only what they wanted to hear. And Lysenko's subordinates quickly learned to give him only the results he wanted. Under his direction, large-scale vernalization of many crops was planned, without any prior research at all. Before the failure of one crop was recorded, other crops were initiated. The positive results he reported were often based on small samples and/or inaccurate or falsified data, and almost always lacked experimental controls.

His "success" created particularly thorny problems for the breeders who applied genetic methods to crop improvement. The small but steady incremental advances produced through breeding would not satisfy the Party bosses, who were demanding immediate dramatic results.

There was escalating tension between Lysenko and his allies who touted immediate "practical" results, and those doing "pure research" on species such as fruit flies. The "problem" with genetics, so it was said, was not merely that genetics might be a distraction from the pressing needs of Soviet agriculture, but that it was invalid and, worse, reactionary and bourgeois—in fact, because genetics implied that there were hereditary constraints imposed on nature and man, it was viewed as contrary to Soviet philosophy.

Lysenko became a Party favorite and had frequent contacts with Party leaders. In a speech in 1935 before agricultural workers at the Kremlin, he declared, "Both within the scientific world and outside it, a class enemy is always an enemy, even if a scientist."

Stalin jumped up clapping and shouted, "Bravo, Comrade Lysenko, bravo!"

Stalin's praise further elevated and emboldened Lysenko, and his attacks on and rejection of genetics escalated. Lysenko's "new theory" of heredity rejected the existence of genes and of self-replicating material.

Confrontations with geneticists became more open, frequent, and hostile. Party leaders joined the chorus, and linked genetics to the growing specter of fascism and Nazism. One prominent commissar dubbed genetics the "handmaiden of Goebbels's department," referring to Hitler's propaganda ministry.

All the while, Lysenko's agriculture programs were disasters. Yields from vernalized wheat were poor, the vegetable supply declined, a potato program was a flop. The crop failures exacerbated the perpetual food shortages. The geneticists struck back, arguing that "if Academician Lysenko were to give greater attention to the principles of modern genetics, it would facilitate his work. . . . While rejecting genetics and the genetic foundation of selection, Academician Lysenko has not given genetics anything theoretically new in exchange"

The Party Central Committee sanctioned an "open" discussion

between the two opposing camps in 1939. Vavilov, Lysenko's former mentor and supporter, pulled no punches: "Lysenko's position not only runs counter to the group of Soviet geneticists, it runs counter to all of modern biology. . . . In the guise of advanced science, we are advised to turn back essentially to obsolete views out of the first half or the middle of the nineteenth century. . . . What we are defending is the result of tremendous creative work, of precise experiments, of Soviet and foreign practice."

Lysenko replied, "I do not recognize Mendelism. . . . I do not consider formal Mendelian–Morganist genetics a science . . ."

Philosopher Pavel Yudin added, "The teaching of genetics must be eliminated from secondary schools."

One month after the meeting, Stalin summoned Vavilov to see him. When Vavilov tried to explain the scientific basis of his work, Stalin turned a deaf ear and dismissed him.

Arrests and Executions

As World War II escalated in the spring of 1940, the USSR faced new and intense agricultural challenges. Somewhat surprisingly, in spite of his sagging support, Vavilov was selected to make an expedition to evaluate and make recommendations on agriculture. It was a ruse to get him out of sight.

While in the Ukraine, a black car arrived at Vavilov's dormitory and four men ushered him into it on the pretense of urgent business in Moscow. It was the NKVD, the secret police. Vavilov was under arrest for "conducting a struggle against the theories and research of Lysenko . . . that possess decisive importance for the agriculture of the USSR."

Lysenko soon forced out all of the geneticists who worked with or under Vavilov. He also purged the biology department of Leningrad University. Many scientists were falsely accused, arrested, and sentenced to be shot. Many others lost their jobs and were forced into other areas of work. Institutes were closed, textbooks were revised. No Soviet scientists attended the International Congress of Genetics of 1939 in Edinburgh, even though Vavilov was its president.

FIG. 9.2. **Nikolai Vavilov.** The internationally renowned botanist (left), and the condemned scientist (right), who died in prison in 1943.

Vavilov, awaiting trial, languished in prison. Only the day before his trial was he permitted to view the charges against him, which included treason, sabotage, espionage, and counterrevolution. The trial took five minutes and he was sentenced to the firing squad.

Vavilov wrote many appeals for mercy, and his sentence was reduced. But he was transferred to a new jail as the German army approached the gates of Moscow. He became emaciated. In January 1943, the most recognized and respected of all Soviet biologists, the youngest person to be elected as a full member of the USSR Academy of Science, former president of the Soviet Geographical Society, recipient of the Lenin Prize, died in prison at age fifty-five.

The pall of "Lysenkoism" continued over Soviet biology for many more years. In 1953 Stalin died, and Nikita Khrushchev took the reins. Of course, 1953 was also the year Watson and Crick discovered the structure of DNA. Surely, this crystalline evidence of the nature of heredity must have taken the wind out of Lysenko and his henchmen?

Or not.

The Denial of DNA

It took the Soviet press three years to finally acknowledge major discoveries in molecular genetics. Translations of articles about DNA were published in journals and science magazines, but Lysenko was unpersuaded: "it is . . . impossible to ascribe an attribute of life, i.e. heredity, to a nonviable substance, deoxyribonucleic acid, for example."

The International Congress of Biochemistry was held in Moscow in August 1961. The crème de la crème of Western biology was going to be there, a great opportunity for Soviet biologists to get back onto the world map. Western scientists assumed that the Lysenko era was over. They were mistaken—many of the Soviet organizers were Lysenkoists.

Zhores Medvedev explains, "They fought against the idea that DNA had a role in heredity! To admit the role of DNA would force them to discard absolutely the inheritance of acquired characteristics."

In an account published in H. F. Judson's *The Eighth Day of Creation*, chemist Vladimir Engelhardt recalled a visit Lysenko made to his institute:

Lysenko says, "All this DNA, DNA! Everybody speaks about it, nobody has seen it!"

I say, "But dear Trofim Denisovich, I could show you a preparation of DNA; it is well known by chemists."

"Show me please."

"Look here, that's the DNA," I say.

Lysenko looks. "Ha! You are speaking nonsense! DNA is an acid. Acid is a liquid. And *that's* a powder. That can't be a DNA!"

The timing of the opening of the congress in Moscow was remarkable with respect to events on the world stage. While the Soviets lagged far behind in biology, they led the space race at the time. Cosmonaut Gherman Titov had just completed the first full day's flight in space and the largest auditorium assigned for the meeting was used for his press conference instead of biology. The most ominous symbol of the East-West divide appeared four days after the congress began, when Khrushchev started building the Berlin Wall.

The biologists would have their own shock when a relatively unknown American biochemist, Marshall Nirenberg, reported the first cracking of part of the genetic code, of how RNA base sequences sequences are decoded into proteins (see chapter 3). It was the highlight of the meeting.

This discovery also gave hope to those seeking to overthrow Lysenko's rule. Medvedev wrote a congratulatory postcard to Nirenberg in early 1962 (the grammar stands uncorrected):

You probably know about long and great discussion in heredity which was started by Lysenko and which seemed to be permanent. The new experimental direct attack on coding problem is a good basis for the settlement of this hard discussion. Therefore, I see to this achievement not only as to the great discovery of natural sciences, but as to the important step which can make influence to the situation in genetics here. . . .

The achievement was completely lost on the Lysenkoists.

After the congress, the political battles continued. Medvedev wrote a *samizdat* book (a secret underground publication) in 1962 detailing the facts surrounding Lysenko's rise and his many failures. The authorities initially would not allow it to be published, but it was widely circulated among scientists.

Medvedev's book was eventually published in the Soviet Union in 1967, but when it was published in the United States in 1969, he lost his job. He was then one of the first dissidents to be put into an insane asylum, for exhibiting signs of "incipient schizophrenia." Under intense pressure from the West, the authorities permitted Medvedev to leave the USSR four years later.

The Lysenko episode and its aftereffects on Soviet biology spanned the duration of the enormous molecular revolution in biology. It is difficult to imagine how the science of genetics and the understanding of DNA could be explicitly denied, and how the failures of Lysenko's ideas could be ignored for so long, by so-called scientists. No amount of evidence could sway the ideologues once they committed to their position. Of course, it is easier to understand why, once the penalties for noncompliance became clear, there was reluctance to challenge the status quo. There were many courageous Soviet biologists who did speak out and paid dearly for their convictions.

Lysenko's impact was felt beyond the borders of the USSR. Mao Zedong, leader of Communist China, imposed similar collectivization schemes on Chinese farmers. Yields fell precipitously, so Mao ordered farmers to use techniques that Lysenko and his cronies advocated, such as close planting, deep plowing, refraining from fertilizers, and extreme pest control. All of these measures backfired and decimated Chinese agriculture. Jasper Becker, in his book *Hungry Ghosts: Mao's Secret Famine*, estimates that as many as 30 to 40 million deaths were caused by the Chinese famine of 1958–1961.

It would be an oversimplification, however, to attribute this tragedy solely to the cultures of China and the Soviet Union and the lack of individual freedoms. In the United States, people are free to think what they wish, yet here, too, "educated" persons with potentially good intentions seem no less prone to devote themselves to fallacious principles.

Spinal Crackers

The concept of chiropractic, a name derived from the Greek *cheiro* ("hand") and *praktika* ("practical" or "operative"), was founded by Daniel David Palmer in the late nineteenth century. A grocer and "magnetic healer" in the American Midwest, Palmer had a patient who lost his hearing in an accident and was not responding to his techniques. He noticed an unusually large bump on the back of the patient's neck, near the fourth cervical vertebra. Palmer reasoned that if the vertebra was put back in position, hearing should be restored. Palmer "racked" the vertebra into position and reported that the man could soon hear. While the claim is a bit astonishing, since the cochlear nerve of the ear does not pass through the neck, this was nonetheless the "eureka" moment for Palmer and the birth of chiropractic.

Palmer developed the theory that nearly all diseases were primarily neurological in origin, and caused by the pinching of nerves by misaligned vertebrae. In Iowa, he founded a school (which is still operational today) to teach his methods of spinal adjustments, and installed his son, Bartlett Joshua, to run it. The younger Palmer advanced his father's ideas about how chiropractic healing works by removing interference with a life force of "innate intelligence" in the body, thereby enabling the body to heal itself.

This founding principle conflicted with the medically accepted view of infectious diseases as being caused by germs. Young Palmer rejected the then prevailing picture of contagious diseases: "Chiropractors have found in every disease that is supposed to be contagious, a cause in the spine." The opposition of chiropractors to the germ theory of disease led to a resolute opposition to vaccination, opposition that continues among a substantial fraction of chiropractors today. One particularly dark and telling episode surrounded the advent of the polio vaccine.

In 1954, the year before the Salk polio vaccine became available, there were 38,476 cases of polio in the United States. In 1955, there were 28,985 cases; in 1956, the figure dropped to 15,140 cases; and by 1961, there were 1312 cases.

As public health authorities ramped up their efforts to encourage vaccination, chiropractors countered with their own campaign. A sample of some communications from various representatives contain some astonishingly ignorant and dangerous statements and claims. Some challenged the efficacy of the vaccine, with the *Journal of the National Chiropractic Association* asking in an article, "Has the Test Tube Fight Against Polio Failed?" Rather than take the vaccine, the article urged that "chiropractic adjustments should be given of the entire spine the first three days of acute polio."

According to the U.S. surgeon general, nine out of ten of the new polio cases occurred among individuals who had not received the vaccine (the Salk vaccine required a series of shots).

Chiropractors claimed successful treatment of both acute and chronic polio, reporting complete recovery in 71 percent of acute cases. Their claims did not take into account that 60 percent of patients who contracted the disease recovered without weakness or paralysis. The patients with more severe polio were treated by conventional doctors with extensive measures, including tracheotomy, feeding tubes, or an iron lung. Chiropractors offered none of these measures, only spinal adjustments.

Chiropractors' lack of training in the diagnosis or causes of disease was no deterrent in discouraging public vaccination. One pamphlet circulated in Colorado stated, "Your thoughtful consideration should be given to the fact that contamination of the human body through the injection of diseased animal cells must be abhorrent to the Creator who equipped us with cell life peculiarly our own, and is not miscible with that of animals."

ONE WOULD think that fifty years after the success of the polio campaign, this nonsense would be behind us. Sadly, it is not.

A 1994 survey of 171 chiropractors reported that one-third believed there was no scientific proof that immunization prevents disease. A 1998 survey of Boston area chiropractors found that only 30 percent actively recommended immunizations, 7 percent recommended against immunization, and 63 percent made no recommendation. And a 2002 survey of Canadian chiropractic students revealed a progressive decline of support for vaccination as students progressed through their years of schooling, with a quarter of the fourth-year students agreeing with the statement that "there is little scientific proof that immunization prevents infectious disease."

Chiropractors practice what they preach. A 1999 survey reported that 42 percent of chiropractors did not have their children vaccinated.

With all of the evidence for the benefit of vaccination against small-pox (now essentially eradicated from the planet), polio, diphtheria, tetanus, mumps, measles, rubella, hepatitis, and more, how can sup-posedly educated persons hold such unreasonable, and for their patients and families, potentially dangerous views? (The first fatal case of childhood diphtheria in the United States in 1998 was the son of a chiropractor who did not believe in immunization).

The best insights into this disconnect between evidence and reason come from chiropractors who have rejected the antivaccination philos-ophy. Two Canadian chiropractors, Jason Busse and Stephen Injeyan, writing in the highly respected medical journal *Pediatrics* with their microbiologist colleague James Campbell, identified several argu-ments and tactics chiropractors and their professional associates have used to oppose immunization. These arguments and tactics are worth study because their flavor greatly resembles those employed by anti-evolution camps. In both cases, there are similar motives of attempt-ing to discredit the science to which they object.

Six of the common arguments or tactics used against vaccination are:

1. *Doubt the science.* Chiropractors have sought other explanations besides vaccine efficacy for the decline of certain diseases. The doubters point to cyclical patterns of disease incidence and sug-gest, for example, that the decline in the incidence of polio was part of a natural cycle. They also point to other measures, such as sanitation and hygiene, as the entire explanation for a decrease in certain diseases. They completely ignore large controlled clinical trials, or attempt to explain away such data as having been manip-ulated, which leads directly to the second angle of attack.

2. *Question the motives and integrity of scientists.* In addition to claims of data manipulation, some doubters have perceived a con-spiracy among scientists and pharmaceutical companies, and implied that their support for vaccination is motivated by greed, not public health.

3. *Magnify disagreements among scientists, and cite gadflies as authorities.* In virtually all scientific arenas, there is room for honest disagreement. In vaccination medicine, the timing and dosing of vaccines, the need to boost later in life, and the risks and benefits of vaccinating those with compromised immune systems (e.g., HIV patients, individuals on chemotherapy, the elderly) are typical legitimate issues. Chiropractors have inflated technical disagreements to imply fundamental disagreements over the value of vaccination. Another tactic is to quote ardent critics who hold medical credentials, no matter how isolated and unsubstantiated their views.

4. *Exaggerate potential harm.* Vaccination, like every other medical procedure, carries some risk that varies with the vaccine and patient population. The incidence of adverse events is well documented and reported, and information concerning these risks is always provided as part of the consent process. Opponents of vaccination often emphasize or exaggerate the risks of vaccination, and typically fail to acknowledge the risk and consequences of not being vaccinated and of acquiring an infection.

5. *Appeal to personal freedom.* Compulsory vaccination of school-age children is viewed by some as an unacceptable violation of personal and parental rights—"a conspiracy aimed at destruction of basic American freedoms," as one Denver chiropractic clinic stated. The Supreme Court has rejected this argument on the grounds that an individual's beliefs cannot subordinate the safety of an entire community.

6. *Acceptance repudiates key philosophy.* Ultimately, once the fog of disinformation clears and the efficacy of vaccines is demonstrated on a massive scale, most of the arguments provide weak refuge. The fallback position is that vaccination is at odds with the so-called major premise of chiropractic. The idea that all diseases stem from misalignment or subluxation of vertebrae is, according to Campbell, Busse, and Injeyan, "taken as a matter of faith that is not open to scientific scrutiny." R. B. Phillips writes in *Journal*

of Chiropractic Humanities: "This faith-based approach negates the need for inductive reasoning with its dependence on probability because absolute truth is already known and only needs personal confirmation through individual observations."

The great and telling irony is that chiropractors who deny the science of vaccination have never put their beliefs to similar double-blind controlled testing. Robert Anderson, an M.D. and a Doctor of Chiropractic, notes that chiropractors "tend to evaluate all things medical in symbolic terms as hostile and harmful" and points out that chiropractors' "belief has not been subjected to testing in clinical trials or laboratory experiments, and thus becomes a matter of belief rather than of scientific verity. A refusal to advocate or submit to vaccines serves conservative chiropractors as an understandable *cultural* [emphasis added] symbol."

Anderson captures the foundation of denial—a belief that is cultural, not a verifiable scientific conclusion.

As Busse, Injeyan, and Anderson demonstrate, not all chiropractors hold such beliefs. But for such a substantial fraction to do so reflects the relative power of cultural ideology over verified science.

And that brings us to evolution.

The Denial of Evolution

The denial of evolution requires far more denial than the modest knowledge of genetics in the era of the Soviet biologists in the 1930s or the knowledge of immunology and virology denied by chiropractors in the 1950s. The denial of evolution requires denial of the bedrock of two centuries of biology and geology. That is quite a feat that demands some explanation.

The denial of evolutionary science has, of course, a long history. My task here is not to recount that history (you will find several recommended books in Sources and Further Reading at the end of this book), but to boil the issues down to their basics in just a few pages. Again, my goal is to equip you with some factual material for discussions that may arise at the dinner table, office, or, perhaps most important, in your or your children's school.

My fundamental premise is that the denial of evolution, like the other instances of denial, is not about the science. It can't be. It is about ideology, in this case religious ideology. While denyers of evolution complain about evidence and market purported scientific challenges to it, these are smoke screens—the very same kind thrown up by the chiropractors who oppose vaccination. We have to pierce those screens to understand the motives behind them. I will rely on a number of direct quotations from evolution's opponents to illustrate these tactics. This is partly a matter of ensuring accuracy and partly to convey the flavor of some of evolution's most ardent critics. I will organize the arguments and tactics into the same six categories as those used by the chiropractors.

1. *Doubt the Science.*

One often encounters blanket statements, such as "virtually no scientific evidence for evolution exists" (T. Bethell, *Christianity Today,* September 3, 2001), "there is no real scientific evidence that evolution is occurring at present or ever occurred in the past. . . . Evolution is not a fact of science, as many claim. In fact, it is not even science at all (H. Morris, "The Scientific Case Against Evolution," *Impact*, no. 330, December 2000), or "evolution is a myth, devoid of any scientific evidence" (P. Fernandes, doctoral dissertation, Institute of Biblical Defense, 1997).

Such conclusions are often presented as logical conclusions to a variety of arguments refuting elements of evolutionary science. Two of the most commonly encountered arguments contain assertions about the absence of "transitional forms" in the fossil record and the role of random mutation. Both arguments are founded on fundamental (perhaps deliberate) misunderstanding of the facts and elements of the evolutionary process.

Paleontology has, in fact, identified many examples of fossils with characteristics that are intermediate between those of different groups. The extensive fossil record of horse evolution, the famous fossil *Archeopteryx* with its bird- and reptilian-like features, feathered dinosaurs, and the early species of four-legged vertebrates are some of the best examples. As the diversity of species known from particular episodes in history has expanded, paleontologists have identified many intermediates in the evolution of key characters. One poignant example of the emptiness of the "no transitional fossils" argument concerns a criticism in 1994 by Dr. Michael Behe, a proponent of the "intelligent design" ideas I will describe shortly, that there were not transitional fossils linking the first fossil whales to their land-dwelling Mesonychid ancestors. Within one year of his statement, three transitional species were identified. While this and many more examples should dispel this tired argument, the objection is still often repeated as though it were true.

Some of the biggest howlers of the anti-evolution argument concern mutations and genetics:

* "Evolutionists need a mechanism that explains how evolution has supposedly occurred. Many evolutionists believe that mutation is this mechanism" (H.M. Morris, *Science and the Bible*).
* "However . . . mutations merely scramble the already existing genetic code. No new genetic information is added" (ibid.).
* "Yet, for evolution to have occurred, a mechanism is needed through which new genes are produced. Therefore, mutations fail to explain evolution. Evolutionists claim that they believe the present interprets the past. However, there is no mechanism that spontaneously produces new genetic information. Until such a mechanism is found, evolution can only be accepted by 'blind faith.'" (P. Fernandes, doctoral dissertation, Institute of Biblical Defense).

Dr. Fernandes apparently got his training in genetics through Henry Morris's *Science and the Bible*, because even a glance at any genetics textbook would have taught him about gene duplication, recombination, insertional mutations, transposition, and translocation—all of which can and do produce new genetic information—not to mention point mutations that can impart new functions, as amply illustrated throughout this book.

The misconception that all mutations are always harmful is repeated so often in the denial literature, it is taken as truth. Yet, biology has known otherwise since the early days of genetics.

Beyond technical issues, a favorite tactic of the anti-evolution camp is to blur the distinctions between "hypothesis," "fact," or "theory" in biology. While in everyday speech, "hypothesis" and "theory" are often equated as merely some form of conjecture, and "facts" as being something more certain, in science "theory" connotes much more. The National Academy of Sciences has defined a scientific theory as "a well-substantiated explanation of some aspect of the natural world that can incorporate facts, laws, inferences, and tested hypotheses." When speaking of evolutionary "theory," scientists are not hedging their support or confidence, as opponents like to imply—it is just a matter of following formal definition.

Pope John Paul II captured the distinction in the context of evolutionary theory in a 1996 statement published in *L'Osservatore Romano*:

> fresh knowledge has led to the recognition that evolution is more than a hypothesis. It is indeed remarkable that this theory has been progressively accepted by researchers, following a series of discoveries in various fields of knowledge. The convergence, neither sought nor fabricated, of the results of work that was conducted independently is in itself a significant argument in favour of this theory.

The Pope's choice of language suggests a clear understanding of the weight of evidence and how consensus is built in the scientific process.

2. *Question the Motives and Integrity of Scientists.*

At the heart of much opposition is the assertion that evolutionary science is motivated by an atheistic philosophy. The most virulent statement I have encountered in recent years was penned by Dr. Ken Cumming of the Institute for Creation Research in response to the Public Broadcasting Service (PBS) television program *Evolution* in September 2001:

> Only thirteen days after the act of terrorism on New York, [PBS] delivered a different, but another event of grave importance that was witnessed by millions of Americans—a seven-part, eight hour special entitled *Evolution*. PBS . . . televised one of the boldest assaults yet upon both our public schools with the millions of innocent school children and the foundational worldview on which our nation was built. These two assaults have similar histories and goals. The public was unaware of the deliberate preparation that was schemed over the past few years leading up to these events. . . . America is being attacked from within through its public schools by a militant religious movement of philosophical naturalists (i.e., atheists) under the guise of secular Darwinism. Both desire to alter the life and thinking of our nation."

I will address this assertion under heading 6 below, with a little help from the Bishop of Oxford.

3. *Magnify Disagreements Among Scientists,*
and Cite Gadflies as Authorities.

Ever since Darwin, biologists have sought to understand the mechanisms of biological evolution and to elucidate the history of life. Every facet of this work has been a matter of testing alternative hypotheses—from the relative contribution of different genetic mechanisms to evolution, to the relationships among species. But the technical disagreements that arise, as they do in any healthy science, must not be mislabeled as disagreement over the fact that evolution has occurred—

that living forms are descended from ancestors and have been modified over the course of time through the process of natural selection.

There are some individuals with scientific credentials who doubt or deny certain elements of evolutionary science that are widely accepted by the scientific community; some may even doubt the entire theory. But getting a doctoral degree and making negative arguments are relatively easy—making new, verifiable discoveries is an altogether different matter. The denyers specialize in rhetoric and the mining of quotes, not in laboratory research.

4. *Exaggerate Potential Harm.*

Opponents of evolution perceive great danger in evolutionary principles and lay much blame for society's difficulties on the influence of "Darwinism" in modern culture. Ken Ham, president and founder of the creationist organization Answers in Genesis, sees a connection between the teaching of evolution and school violence:

> Evolutionary indoctrination declares that we are only animals in the struggle for survival. This has created a mindset in many of our young people that life lacks purpose. . . . Some students—who are brainwashed in evolution—believe that life is all about death, violence, and bloodshed because, after all, these are the processes by which they evolved.

Darwin has also been blamed for Soviet-style communism. Cal Thomas, a syndicated columnist, reacted to the Pope's 1996 statement on evolution by alleging that the Pope "has accepted a philosophy that stands at the core of communism. Why would he want to accept the heart of a worldview that he spent his life opposing?" This is a very curious statement in light of Lysenko's antigenetic stance and rejection of Darwinian evolution.

An association between the Holocaust and Darwin's understanding of the competitive struggle for survival is another often repeated allegation. Jerry Bergman writes, "of the many factors that produced the

Nazi holocaust and World War II, one of the most important was Darwin's notion that evolutionary progress occurs mainly as a result of the elimination of the weak in the struggle for survival." He concludes, "If Darwinism is true, Hitler was our savior and we have crucified him. . . . If Darwinism is not true, what Hitler attempted to do must be ranked with the most heinous crimes of history and Darwin as the father of one of the most destructive philosophies of history."

By inflating evolutionary science into a political philosophy, Bergman and others seek to discredit the science. Geneticist and author Steve Jones has referred to the hijacking of Darwin's theory into political causes as "vulgar Darwinism" and notes, "Evolution is a political sofa that molds itself to the buttocks of the last to sit upon it."

5. *Appeal to Personal Freedom.*

The teaching of evolution in public schools is frequently viewed as an assault upon the religious freedom of those who oppose it. Those holding this view have advocated placing disclaimers in textbooks that raise doubt about the claims of evolutionary science, and they have appealed for the teaching of "alternative" views of the history of life. The latter generally involves those alternatives embracing some version of creationism and is often justified on the basis of a sense of "fairness" or "balance." Both tactics have been repeatedly struck down as unconstitutional in federal courts.

For example, a federal judge in Atlanta, Georgia, recently ruled that stickers placed on biology textbooks in the Cobb County school district (figure 9.3) violated the "establishment clause" of the First Amendment of the U.S. Constitution, which states that "Congress shall make no law respecting an establishment of religion, or prohibiting the free exercise thereof," a prohibition that applies to states through the Fourteenth Amendment. The judge's decision relied on an extensive body of case law in which the Supreme Court and lower federal courts have struck down anti-evolution statutes, policies, and disclaimers, as well as balanced-treatment legislation.

The judge found in particular that the statement "Evolution is a

This textbook contains material on evolution. Evolution
is a theory, not a fact, regarding the origin of living things.
This material should be approached with an open mind,
studied carefully, and critically considered

Approved by
Cobb County Board of Education
Thursday, March 28, 2002

FIG. 9.3. **Evolution disclaimer on Cobb County, Georgia, textbooks.** The stickers were ordered removed by Federal Judge
Clarence Cooper on January 13, 2005.

theory, not a fact, regarding the origin of living things" ran afoul of
the establishment clause. Furthermore, he wrote that the sticker "mis-
leads students regarding the significance and value of evolution in the
scientific community for the benefit of religious alternatives" and that
"the sticker targets only evolution to be approached with an open
mind, carefully studied and critically considered without explaining
why it is *the only theory being isolated as such* [emphasis added]."

Despite the extensive and growing body of federal case law, school
districts in many states are actively considering "balanced treatment"
and other policies that should also run into constitutional problems.

6. Acceptance Repudiates Key Philosophy.
The ultimate source of the conflict over evolutionary science is the
same that we have seen over genetics in the Soviet Union and vaccination in the conservative chiropractic community—it is viewed to be at
odds with matters of faith that are not open to scientific evidence. As
stated by the organization Answers in Genesis, "the real issue [is] the
authority of the Bible as a trustworthy revelation from God, and
hence the integrity of its Gospel message."

David Cloud of the Way of Life Fundamental Baptist Information Service argues that there are three reasons to reject evolution:

1. "First, we must reject evolution because it denies the Bible," particularly statements in Genesis. Cloud suggests that "if the Bible does not mean what it says, there is no way to know what it does mean."
2. "We must reject evolution because it denies God." Cloud argues that the "God of the Bible is involved with every detail of creation" and that "the god of evolution is not the God of the Bible."
3. "We must reject evolution because it denies salvation." Cloud writes that "if Genesis 1–3 is not literal history, the rest of the Bible and the doctrine of Christ and salvation is a fairy tale—because it is all based on a supposition that these were real historical events."

I believe this is a fair representation of the concerns and convictions of a majority of evolution's opponents and denyers. Similar concerns are echoed by many prominent members of that contingent. Ken Ham believes that "until our nation allows God to be the absolute authority, and accepts the Bible as truth," all manner of difficulties will continue to proliferate.

But while these most vocal and politically active opponents of evolution have defined the positions of Christianity and evolution as an either/or choice, and the popular press appears to promulgate that view, it is clearly not correct. The conclusions concerning the irreconcilability of Christianity and evolutionary science are not shared by many scientists, theologians, leading clergy, or even entire Christian denominations.

Consider, for example, a recent statement by Reverend Richard Harries, the Bishop of Oxford; in a BBC radio "Thought for the Day," he expressed his sadness and concerns over the treatment of evolution:

Do some people really think that the worldwide scientific community is engaged in a massive conspiracy to hoodwink the rest of us?

. . . The theory of evolution, far from undermining faith, deepens it. This was quickly seen by Frederick Temple, later Archbishop of Canterbury, who said that God doesn't just make the world, he does something even more wonderful, he makes the world make itself. . . . The second reason I feel sad about this attempt to see the Book of Genesis as a rival to scientific truth is that it stops people taking the Bible seriously. The Bible is a collection of books made up of very different kinds of literature, poetry, history, ethics, law, myth, theology, wise sayings and so on. . . . The Bible brings us precious, essential truths about who we are and what we might become. But biblical literalism hinders people from seeing and responding to these truths . . . biblical literalism brings not only the Bible but Christianity itself into disrepute.

The bishop has a lot of distinguished company. The General Assembly of the Presbyterian Church reaffirmed in 2002 that "there is no contradiction between an evolutionary theory of human origins and the doctrine of God as Creator." Similarly, the United Church Board for Homeland Ministries stated in 1992, "The assumption that the Bible contains scientific data about origins misreads a literature which emerged in a pre-scientific age."

Thus, the fundamentalist Protestant position that insists on the rejection of evolution and the literal truth of the Bible in all matters, including science, is at odds with not only two centuries of modern science, but established doctrines and positions of the Catholic, Jewish, and a broad array of Protestant faiths. Furthermore, the fundamentalist position ignores the fact that a significant proportion of scientists also hold mainstream religious views.

The principal means for evolution's deniers to circumvent the vast body of evolutionary science has been to employ the tactics of rhetoric and propaganda I have described above, which has succeeded in promoting the idea that evolutionary science is at odds with *all* denominations. The second major effort has been aimed at influencing the political process at local, state, and national levels. Given that all of the opponents and organizations I have cited have overtly religious aims and affiliations, this would seem to be a difficult maneuver in light of the establishment clause of the U.S. Constitution. This impediment has given birth to a new strategy—the attempt to wrap religious belief in the skin of scientific credibility.

Old Wine in a New Bottle: The Myth of Intelligent Design

The "newest" incarnation of alternatives to Darwinian evolution is dubbed intelligent design. The major premise of this movement is that some biological structures are too complex to have evolved in a stepwise fashion under natural selection, and therefore, they must have been "designed" by an Intelligent Designer. Who is this Designer? Well, proponents often take pains not to name Him or Her out loud, but God would be a good guess.

The intelligent design idea traces back to William Paley, who articulated the general concept in his *Natural Theology* 200 years ago. Simply put, the argument is that just as whenever one sees intricacy and complexity in man-made objects, one infers a designer, so, too, with the contrivances of Nature.

Interestingly, one of the most recognized proponents of intelligent design, Dr. Michael Behe, accepts the antiquity of the Earth, the descent of species from common ancestors, and the origin of some traits by Darwinian natural selection. His major thesis is that some structures and systems in organisms are "irreducibly complex" and supposedly nonfunctional if any of their parts are removed. These systems and structures could not, in Behe's view, have arisen by natural selection.

Behe's acceptance of many facts of biology puts him in an awkward position in terms of explaining the when and how of intelligent design. The material in this book on the interaction of mutation, selection, and time on the evolution of the text of DNA presents a stark contrast and devastating evidence against Behe's notions. In his book *Darwin's Black Box*, Behe posits, "Suppose that nearly four billion years ago the designer made the first cell, already containing all of the irreducibly complex biochemical systems discussed here and many others. (One can postulate that the design for systems that were to be used later, such as blood clotting, were present but not 'turned on.' In present-day organisms plenty of genes are turned off for a while, sometimes for generations, to be turned on at a later time)."

This is utter nonsense that disregards fundamentals of genetics. Dr. Ken Miller of Brown University has described this scenario as "an absolutely hopeless genetic fantasy of 'pre-formed' genes waiting for the organisms that might need them to gradually appear." As we saw in chapter 5, the rule of DNA code is use it or lose it. The constant bombardment of mutation will erode the text of genes that are not used, as it has in icefish, yeast, humans, and virtually every other species. There is no mechanism for genes to be preserved while awaiting the need for them to arise. Rather, as we saw in chapter 4, gene duplication, an observable process in living species, is just one means available for expanding genetic information and complexity as life evolves, and the signature of gene duplication events throughout evolution abounds in the DNA record.

As just one example, let's consider the distribution of β-globin genes that encode one of two chains of the hemoglobin molecule. We humans have five β-globin genes and they are all located right next to one another on chromosome 11. Different genes are used at different times of life with the so-called epsilon (ε) genes active in embryonic life, the two gamma (γ) genes active in fetal life, the delta (δ) globin expressed at low levels in adults, and the β-globin most highly expressed in adults. Chickens have four β-globin genes; most fish have fewer β-globin genes, which are arranged differently than those in birds and humans—except for the icefish, which has no functional β-globin genes at all.

What could account for this pattern of β-globin gene distribution? Under the intelligent design preformed gene model, all of the genes we use would have been designed long, long before humans or other mammals evolved to use them. If so, then why would there be fewer β-globin genes in other groups of vertebrates, and none at all in the ice-fish (a more recently evolved group of fish)? Shouldn't those silent preformed genes still be there in other species, waiting to be called upon? And how did the icefish wind up with no globin genes, and a fragment of an α-globin gene? What Designer designs partial, non-functional genes?

The evolutionary explanation of the distribution of globin genes is much simpler. Vertebrate ancestors had fewer globin genes in a fish-like arrangement, and the duplication of these genes, followed by their divergence through mutation, produced the large, diverse set of genes used during our lifetimes. This is a pattern seen in hundreds of gene families throughout evolution.

Advocates of intelligent design complain that the concept gets no respect in the scientific community because of some prejudice. Not so. We like new hypotheses, but we strongly prefer those that are consistent with well-established facts. As a result, intelligent design does not rise anywhere near the level of *theory*. It has produced no insights into any scientific question and it is inconsistent with rigorously tested knowledge. Were it not for its theological appeal and the tactics of its proponents, we would have never heard of intelligent design, and it would join other notions in the vast ashcan of rejected ideas.

At best, then, intelligent design is a myth. There are many myths about creation and the workings of the natural world. I am particularly enchanted with those of Australia's aborigines. The myth of the Dreamtime and the characters depicted in their remarkable tradition of rock art are wonderful elements of their culture. But the Dreamtime is not science, so we do not teach it in science class—nor any other myths about origins or natural phenomena, and certainly not as *alternatives* to established science.

Fortunately, a U.S. federal court agrees with this assessment. In the most important legal test of intelligent design to date, Judge John E. Jones III of the United States Distinct Court for the Middle District of

Pennsylvania ruled on December 20, 2005, that the Dover, Pennsylvania, school board trampled the U.S. Constitution when it invoked a policy that put forth intelligent design as an alternative to evolutionary theory. In a devastating autopsy of intelligent design, the judge found overwhelming evidence that intelligent design "is a religious view, a mere relabelling of creationism, and not a scientific theory," with "utterly no place in a science curriculum." Judge Jones also underscored that "many leading proponents of [intelligent design] make a bedrock assumption which is utterly false . . . that evolutionary theory is antithetical to a belief in the existence of a supreme being and to religion in general."

In late 2004, encouraged by members of the intelligent design movement, the school board in Grantsburg, Wisconsin (my home state) considered the teaching of "alternative theories" of origins. The Grantsburg school board promptly received a letter signed by a few hundred scientists, urging them to abandon this course. Then they received a similar letter from religious studies professors. But the most persuasive letter, and one that symbolizes some beacon of hope, was signed by 188 pastors of Baptist, Catholic, Lutheran, Methodist, Episcopal, and other churches across the state (they have since been joined by more than ten thousand clergy across the United States). I will give them the last word on this matter; their letter read in part: "We believe that the theory of evolution is a foundational scientific truth, one that has stood up to rigorous scrutiny and upon which much of human knowledge and achievement rest. To reject this truth or to treat it as 'one theory among others' is to deliberately embrace scientific ignorance and to transmit such ignorance to our children."

Amen.

Why Does Evolution Matter?

Understanding and accepting evolution is a matter of adhering to the scientific process, which has delivered enormous advances in agriculture, medicine, and technology. Just as the science of DNA has permeated our daily lives in forensic analysis, paternity testing, and the diagnosis, prevention, and treatment of disease, so, too, should its meaning for our understanding the real history of life and our species. And, just as the foundations of paleontology are underpinned by vast amounts of geological knowledge, the foundations of this new DNA record of evolution are underpinned by vast amounts of knowledge in cellular and molecular biology, genetics, embryology, and physiology.

The facts of astronomy, microbiology, and genetics were resisted by certain groups until the tangible, visible evidence was overwhelming. The tangible record of evolution in DNA is overwhelming, and cannot be argued away. Those who have continued to oppose evolution in the face of the evidence cannot be allowed—like the incredulous French doctors, dictators of Soviet biology, or fundamentalist chiropractors— to silence or ignore science in order to accommodate their disbelief.

When the scientific process is abandoned, the lesson throughout history is failure or outright disaster in human affairs. In the next and final chapter, I will focus on the significance of evolution and the importance of adhering to the scientific process in our responsibility as the stewards of our planet.

Fossil palm tree and fish from the Eocene epoch, approximately 50 million years ago, at Fossil Butte, southwestern Wyoming.
Photograph courtesy of Shirley Ulrich, Ulrich's Fossil Gallery, Kemmerer, Wyoming.

Chapter 10

The Palm Trees
of Wyoming

· · · · · · · · · · · ·

The farther backward you can look, the farther forward
you are likely to see.

—Winston Churchill

"THE GREATEST WONDER OF THE NINETEENTH
century!" boasted *The Pacific Tourist* in 1876.

Construction of the transcontinental railroad had begun
in earnest shortly after the Civil War. The lines of two rival
railroad companies, the Central Pacific building from the
west and the Union Pacific building from the east, were
extended at breakneck speed and finally met at
Promontory, Utah on May 10, 1869, completing the route at
the then gigantic cost of over $120 million.

The true heroes of this great achievement were the workers who laid the "iron road" with their bare hands. The
building of the Union Pacific Railroad alone required
300,000 tons of iron rails and more than 23 million spikes.
Beyond the prodigious physical demands of twelve-hour
shifts in searing summer heat, there were great dangers.
Premature explosions of black powder and nitroglycerine,

FIG. 10.1. **The "petrified fish cut" of the Union Pacific Railroad.**
The rock layers contain rich horizons of Eocene fossils. *Photograph*
by Andrew Joseph Russell (1869), courtesy of the Sweetwater County
Museum, Wyoming.

rockfalls, and constant attacks from unsubdued Native American
tribes, who understood what the coming of the railroad signified for
their way of life, all plagued the workers.

But there was also a sense of greatness in the endeavor, and the adven-
ture of encountering and conquering the vast, unknown, and the
sparsely populated landscape of the American West. Setting out to the
west from Omaha, the Black Hills, the Wyoming Basin, the Wasatch
Range, and other great formations and wonders lay ahead.

In 1868, two employees of the Union Pacific, A. W. Hilliard and
L. E. Rickseeker, were surveying about two miles west of Green River,
Wyoming (then part of the Dakota Territory). They blasted a cut
through the barren rock formations typical of the area. Within the
slabs of exposed shale they discovered enormous quantities of well-

preserved fossil fish and plants. They did not know it, but these fossils were part of one of the largest, best preserved fossil beds anywhere on the planet.

The greatest wonder of the nineteenth century was going to pass right through the greatest wonder of the Eocene epoch—laid down some 500,000 centuries earlier.

The workers turned the fossil fish over to Ferdinand V. Hayden, who was conducting a survey of the territory for the U.S. Department of the Interior. Hayden, a surgeon by training, once remarked "my love for Natural History is so great that I hardly feel any disposition for anything else." He was well-known to the Sioux, who called him "he who picks up rocks running" and apparently considered him odd, but harmless. Following the railroad workers' discovery, Hayden published the first descriptions of the fossil-rich formations of the Green River. This was not his only major contribution to science. Hayden collected the first American dinosaur fossils in 1869 and his work in the territory helped lead to the establishment of Yellowstone National Park in 1872.

The "petrified fish cut" was part of a much larger deposit formed by an ancient lake bed. In addition to the several species of herring-type fish found in the rock, sometimes numbering in the hundreds in a single square yard (figure 10.2), there were spectacular stingray, paddlefish, and gar fossils, along with crocodile, turtle, bird, bat, and small horse fossils.

And there were also, of all things, magnificent palm trees—specimens up to ten feet tall preserved in exquisite detail.

Hayden correctly concluded that what is now a semiarid desert of flat-topped buttes and badlands was once a lush tropical forest, with a climate and flora much like the southeastern United States today. The fossil-rich bands of shale date from about 40 to 50 million years ago and now lie above the carved valleys of southwestern Wyoming. Within several layers (called horizons) in the 200- to 300-foot-thick Green River formation, the exceptional fossils abound. If you have ever seen a well-preserved fossil fish set in a light tan-colored rock, chances are that it was quarried out of the Green River formation.

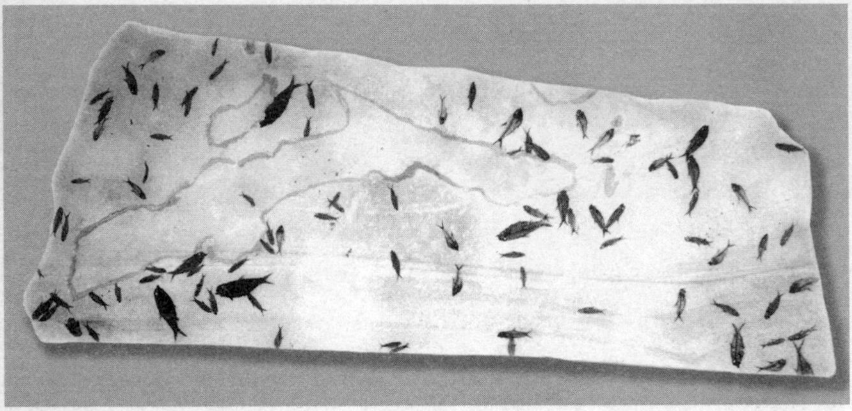

FIG. 10.2. **Massive fish kill.** The massive die-off is preserved in a single fossil layer in this slab from the Fossil Butte area. *Photograph by Jamie Carroll; slab courtesy Ulrich's Fossil Gallery, Kemmerer, Wyoming.*

Part of this vast Fossil Lake formation was set aside in 1972 as Fossil Butte National Monument.

The magic of Fossil Butte stems from both the exceptional beauty and quality of its fossils and the stark contrast between the habitat they represent and the area today. The fossil palm trees and crocodiles remind us how much conditions in any locale do change over time, and with those changes, how species vanish and others replace them. It reminds us how the "fittest" is a conditional, if not precarious, status.

Fossil Lake was the smallest of three large lakes that once covered parts of Wyoming, Colorado, and Utah. Fifty miles long and twenty miles wide at its maximum, the lake existed for about 2 million years, a long time for almost any lake, but profound climate changes eventually led to its disappearance. These climate changes dramatically altered the conditions of selection and the flora and fauna.

I HAVE focused to this point in the book on the making of the fittest. In this final chapter, I will focus on their unmaking—that is, the

FIG. 10.3. **Fossil Butte landscape.** What was once a lush tropical forest is now a semiarid, wind-blown, eroded landscape. *Photograph
courtesy of Arvid Aase, provided by the United States National Park
Service.*

decline of species. The record of life's history is paved with creatures
that were once abundant—trilobites, dinosaurs, and ammonites, to
name a few—but which are now completely vanished due to natural
causes. Here, I will discuss how human activities have impacted or are
currently affecting many once very abundant species, at a far more
rapid rate than anything we know about from natural history. This
"unnatural" mode of selection by humans has many unintended consequences. I will focus in particular on the world's fisheries, which are
gravely threatened by a gathering storm of overfishing, environmental
degradation, and climate change.

Almost fifty years ago, evolutionary biologist Julian Huxley, grand-
son of the great biologist and Darwin's key ally Thomas Huxley, and
brother of the novelist Aldous Huxley wrote in *New Bottles for New
Wine*, "It is as if man had been appointed managing director of the

biggest business of all, the business of evolution . . . whether he is conscious of what he is doing or not, he is in point of fact determining the future direction of evolution of this earth. That is his inescapable destiny, and the sooner he realizes it and starts believing it, the better for all concerned."

Denial or ignorance of our effects on evolution, combined with the politics of self-interest, has already clobbered several species key to the economy, as well as and the livelihoods of hundreds of thousands of people who depended on them. Extreme danger signs are flashing for many more species experiencing selection intensities that are greater than those to which the populations can adapt. We will see how knowledge of evolutionary processes is not merely some aesthetic or philosophical matter, it is the foundation of sustainable policies.

Unnatural Selection

The bighorn sheep is perhaps the most majestic symbol of the high country of Wyoming and surrounding ranges. While tourists hope for a glimpse of these often elusive mammals, hunters prize world-class trophy rams like few other legal quarry. Hunting permits have fetched into the hundreds of thousands of dollars at auction. The income from a few selected hunts is used to support conservation measures, the general thinking being that the taking of a few rams is balanced by the overall economic benefit to wildlife management.

However, long-term studies have revealed some critical unintended consequences of the selective hunting of trophy rams. Hunters most desire rams with the largest horns. Unfortunately, so do female bighorns. In bighorn sheep, most of the growth of the horn length occurs between the ages of two and four. Mating success increases with dominance rank and horn length, and increases from about age six onward. But most rams are shot before reaching age eight, and some are taken as young as age four.

Over a span of thirty years, rams at one locale in the Canadian

FIG. 10.4. **Bighorn ram.** Horn length and body size greatly influence mating success in this species. *Photograph by Don Getty.*

Rockies have shown a marked downward trend in their "breeding value," indicated by a decline in mean body size and horn length. These characteristics are determined to a significant degree by heredity. The effect of hunting has been to select against the faster growing, more robust rams, and therefore favor smaller bodied, shorter-horned rams. The form and genetic makeup of the sheep population are evolving away from their naturally selected optima. It is everyday math again at work—that math of variation and selection over time—only in this case in the wrong direction.

The evolution of the bighorn sheep population highlights how human selection for particular traits can drive evolution in the opposite direction from what we desire, or what is best for natural populations. This is one of the major problems in the management of many natural resources, particularly fish. Without evolutionarily enlightened management, guided by good data, the tragic episodes I am about to describe are bound to be repeated, with catastrophic, irreversible consequences.

Cape Codless

One of the greatest impediments facing early oceangoing explorers was the availability of food. In order to travel long distances, methods for preserving and storing provisions were crucial. The development of techniques for drying and salting cod made longer voyages possible, and contributed to the discovery of Greenland, Iceland, and North America. In 1497, John Cabot (né Giovanni Caboto) set out to find the sea route to Asia that Christopher Columbus missed. He landed instead at a place he found ideal for salting and drying fish and named it "New Found Land." Vessels from all over Europe soon converged on the seemingly limitless populations of cod in this new fishery and within forty years 60 percent of all fish eaten in Europe originated from the North Atlantic coast.

In 1602, Englishman Bartholomew Gosnold found a place he called Pallavisino, so rich with cod he reported being pestered by the fish (we now call this place Cape Cod). Cod fishing became a central way of life in the new colonies. In towns like Gloucester, Massachusetts (on Cape Ann), it dominated commerce. While Gloucester had a population of only about 15,000 residents for most of the nineteenth century, it lost 3800 men to the sea.

On July 2, 1992, after 500 years of operation, the Canadian Grand Banks Northern Cod fishery was closed due to the total collapse of the fish population. The population of northern cod had dropped 99.9 percent from the level of that in the 1960s. Twenty thousand Canadian fishermen lost their livelihoods.

In Gloucester, some fishermen saw the same fate coming to them.

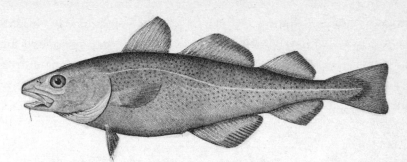

FIG. 10.5. **The Atlantic cod.** Once cod were so abundant that early
sea captains complained of being pestered by the fish. The northern
population has been decimated by overfishing.

Catches had dwindled to a fraction of levels of only ten or twenty
years earlier. The federal government started buying up boats in order
to encourage fishermen to abandon their traditional work, and parts
of the offshore fishery were closed in order to help the fish populations recover.

The cod fishery has not recovered.

What happened?

In short, the cod and the fishermen that depended on them are victims of that universal recipe of variation and selection. Cod fishermen
selected the largest (and oldest) fish. Therefore, fish that, due to genetic
factors, tend to mature at a larger size and greater age were more likely
to be caught before they could reproduce. The removal of these fish
would then favor smaller-maturing fish. With extensive fishing, as the
population size declined, fish matured earlier and at smaller sizes.

With the fish maturing more rapidly, one might think that the population could then rebound more quickly. That might be true in a fish
tank but not in the wild ocean, where other species prowl. Where cod
was once a dominant predator on bottom-dwelling crustaceans and
pelagic fish, these latter animals experienced a population boon.
Species that were formerly cod prey became cod predators, and they

consumed cod eggs and larvae, stunting the recovery of the cod population. Remember that the "fittest" is a matter of both reproduction *and* survival. The smaller cod are not doing well in an ecosystem that has been turned upside down. This consequence of overfishing and imbalance in ecosystems is not at all restricted to cod, but has been observed for many large ocean predators.

Where Did All the Big Fish Go?

In addition to cod, other large fish in the sea include tunas, swordfish, marlin, sharks, rays, and a host of "groundfishes" (flounders, halibut, and skates). Many species of large fish are commercially significant, with some of the tunas being the most prized delicacies.

After World War II, fishing for species became industrialized, where big "factory" ships efficiently scoured vast tracks of ocean for large fish. Based upon a extensive analysis of data from trawler and longline catches from 1952 to 1999, biologists Ransom Myers and Boris Worm of Dalhousie University in Halifax, Nova Scotia, concluded that the populations of large predatory fish today are about 10 percent of preindustrialized levels.

In four continental shelf and nine oceanic systems, Myers and Worm found the same trend: populations dropped by approximately 80 percent within the first fifteen years of exploitation (figure 10.6). On longlines, catch rates fell from 6–12 fish per 100 hooks to 0.5–2 fish per 100 hooks within the first ten years of fishing. This drop in "catch per unit effort" is actually worse than it appears, because the fish now being caught are also dramatically smaller. Myers notes that "the average size of these top predators is only one-fifth to one-half of what it used to be. The few blue marlin today reach one-fifth of the weight they once had. In many cases, the fish caught today are under such intense fishing pressure, they never even have the chance to reproduce."

Myers has been one of the most prolific and vocal researchers trying to correct the widespread notion that the seas are limitless

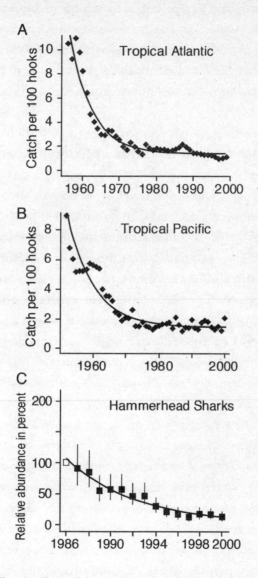

Fig. 10.6. **The decline of large fish populations.** The number of large fish has declined in both the Atlantic (A) and Pacific (B) oceans as measured by longline catch rates (diamonds and boxes). Hammerhead sharks show a decline typical of most shark species (C). *Data is from R. Myers and B. Worm (2003),* Nature *423:280, and J. K. Baum et al. (2003),* Science *299:389; redrawn by Leanne Olds.*

resources and that marine species are extinction-proof. As late as 1883, even Thomas Huxley believed that fish stocks were inexhaustible. Myers writes that the main reasons behind these ideas were "the seemingly inexhaustible abundance of marine life, the remoteness of many marine habitats and the perceived high fecundity of marine fish populations. All of these arguments have been shown to be wrong."

We humans may have little concern for the livelihoods of hammerhead or white sharks, but in another study (of which Myers and Worm were co-authors) it was found that their populations had declined by 75 percent, and all shark populations by 50 percent, over the past fifteen years. Some of this decline is due to so-called by-catch, the snaring of species on hooks or in trawls intended for other species. As top predators, sharks have important influences on the structure of food webs in the ocean, and on the functioning of ecosystems. The reductions of shark populations thus changes the conditions of selection in ecosystems, with some unpredictable consequences.

Trawling imposes another dimension of selection—on fish size. For example, the regulated mesh size on commercial fishing gear over the past several decades has ranged from 7 to 14 centimeters. This means that some species are inadvertently caught in trawls while others slip through. The impact of this size selection is well documented for skates in the North Atlantic.

Over the past several decades two small skate species, the thorny skate and smooth skate, have actually increased in numbers. However, the barndoor skate, an animal that reaches up to 1 meter in width and is about 20 centimeters long at hatching, is near extinction. The decline of the barndoor skate went unnoticed for years, until Jill Casey at Memorial University in Newfoundland and Ransom Myers discovered that not a single animal had appeared in trawls off southern Newfoundland in the last twenty years. Forty-five years ago, barndoor skates were caught in 10 percent of all trawls. The only recent finding of the skate was at depths of 1000 meters offshore, in an area being considered as a new halibut fishery.

The decline of large predators, or the selective loss of other large fish, cannot be viewed simply as losses of single species. It is crucial to understand the consequences of overfishing and by-catch in the ecosystems in which they live (or lived).

Dominoes

There are several types of natural coastal marine habitats in the world, including coral reefs, kelp forests, sea-grass beds, and estuaries. In each of these communities, certain types of organisms are key to their structure and diversity, and complex webs of interaction exist between members. Overfishing and other human activities disrupt these webs, with often catastrophic effects.

For example, kelp forests provide home to many fish, invertebrates, and certain mammals such as sea otters. Sea urchins graze on the kelp, and the urchins are in turn food for the otters and groundfish such as cod. Where the cod have been eliminated by overfishing, or the otters killed off by fur hunters, the sea urchins have overrun the kelp and "deforested" the area. This is the domino effect of removing top consumers in the food web. If sea urchins are then harvested by fishing, the kelp returns, but the other members of the community are still missing, so the kelp forest is permanently impoverished.

Coral reefs are also home to diverse species of fish and invertebrates that are in turn preyed upon by larger animals. The crown of thorns feeds on corals and appears to be held in check by certain species of fish. In the 1980s, massive outbreaks of these starfish on the Great Barrier Reef led to a dramatic decline in coral populations, robbing many animals of their only habitat.

The health of the sea-grass beds that cover various bays, lagoons, and other coastal areas is similarly dependent on the animals that forage on them, and the animals in turn depend on the sea grass. Sea turtles crop sea grass, a process that helps keep nutrients in the food chain and out of sea-bottom sediments. Turtle populations have been dramatically reduced or exterminated in most locales, which has contributed to the die-off and decline of sea-grass beds.

Dugongs are also important tenders of sea-grass beds. Australian colonists reported vast herd of dugongs in Moreton Bay on the Queensland coast in the late nineteenth century. Three- or four-mile-long herds numbering tens of thousands of animals have since been reduced to a total population of just 500 individuals. Sought for their oil and flesh, the dugong "fishery" was rapidly exhausted in short order.

Jeremy Jackson of the Scripps Institution of Oceanography in La Jolla, California, and eighteen marine scientist colleagues have analyzed the vast changes in coastal ecosystems since the dawn of civilization approximately 10,000 years ago. They see a similar pattern of decline in three stages across the world. First, there was limited aboriginal use. Then, increased exploitation over the period of European expansion and colonization. And, finally the recent rapid exhaustion of populations and ecosystems. They note that the baselines of marine populations of 1950 or 1960 used to calibrate recent change are themselves but a fraction of populations before humans began exploiting them.

Jackson and colleagues caution that "few modern ecological studies take into account the former natural abundances of large marine vertebrates" and remind us that so many places in the world—islands, towns, bays, etc.—are named after animals that either no longer exist there or are but a remnant of once massive populations. The tens of thousands of sea turtles that visit certain nesting areas may seem like a great horde, until one considers that these populations numbered tens of *millions* a few centuries ago.

A Perfect Storm

While overfishing is no doubt to blame for the massive reduction of natural populations, the growth in human populations and industry since the Industrial Revolution has introduced additional stresses on already depleted ecosystems. A perfect storm is brewing—of overfishing, pollution, and man-made climate change—that threatens to extinguish ecosystems beyond any chance of recovery. One need only look at habitats close to population centers for abundant evidence of the synergistic effects of these three forces.

Consider, for example, Chesapeake Bay, which is the largest estuarine bay in the United States. More than 150 major rivers and streams drain into the 200-mile-long bay. When Captain John Smith explored the bay in the early 1600s, he described a clear body of water, with meadows of crab-filled sea grasses, oyster reefs so enormous that they were a hazard to navigation, and shoals of rockfish and other species.

Among all marine habitats, temperate estuaries have suffered the greatest damage. The vast oyster reefs that once populated Chesapeake Bay may have filtered the equivalent of the entire water column every three days. However, once industrial oystering began with the use of dredges, the oyster population was decimated. And, as we have seen, the removal of species that control the structure of ecological communities has profound consequences.

After the oyster fishery collapse, symptoms of the bay's weakened condition appeared. Oxygen levels dropped and outbreaks of disease occurred. The latter were clearly exacerbated by a second new ingredient of intense selection on natural populations—pollution from human activities. The damaged ecosystem was unable to handle changes in sediment, nutrient, and microbe levels brought on by industrial activities and farming. Despite some twenty years of attention from government and private agencies, in 2004 a suffocating "dead zone" of oxygen-depleted water covered more than 35 percent of the bay's main stem.

The same pattern of damage, decline, and decay of estuaries once dominated by oyster reefs has occurred elsewhere in the world. The repetition of evolution is, unfortunately, too predictable in these scenarios.

Now, in addition to overfishing and pollution, a third new ingredient of the intense selection on natural populations has appeared—climate change. The quantitative impacts of climate change are more difficult to assess because of the already degraded state of so many ecosystems. Nevertheless, it is certain that local changes in climate have profound effects on ecological communities. Communities that are already over-stressed and overfished have one more force to manage.

Repairing the damage from overfishing and pollution in any locale is a substantial challenge. Doing so against a backdrop of further global changes is even more daunting. With such large human popula-tions surrounding the overfished Atlantic and Pacific oceans, one might think that the vast Southern Ocean and the unpopulated Antarctic might be the one remaining refuge for marine biodiversity. Sadly, that is not the case and the endless repetition of history that led to the disappearance of cod, swordfish, marlin, dugongs, oysters, tur-tles, sea otters, and more from northern waters has been and is being replayed in the Southern Ocean.

Return to Bouvet Island

The main reason Ditlef Rustad and Johan Ruud found themselves in the Southern Ocean was that declining whale stocks in northern waters forced Norway's whaling fleet to look elsewhere. Norway began whaling in the Antarctic's waters in the early 1900s, taking 236 whales between 1904 and 1906. This figure rose to 10,760 whales by 1912, and to 40,000 by 1940.

Whalers pursued one species after another as populations declined. For example, in 1930–31, 29,000 blue whales were taken. The hunters moved on to the fin whale, the sei whale, humpback whales, and then minke whales. By the time most countries agreed to a moratorium on Antarctic whaling, the former population of 200,000 blue whales had been reduced to just 6000 animals, the population of humpback whales was also reduced to the same small fraction, and the fin and sei whales were reduced to 20 percent of their original numbers.

Rustad's primary research was on krill, the short crustacean at the center of the Antarctic food web. As the whaling industry faded, krill

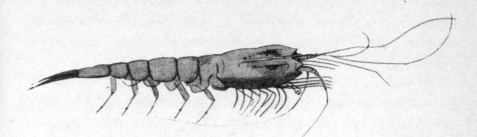

FIG. 10.7. **The krill, *Euphasia superba*.** This shrimp may be the most abundant animal on Earth and is critical to the Antarctic food web. Its density has declined by 80 percent over the past seventy-five years.

became a target for large-scale harvesting, beginning in 1972–73. Krill is processed into products for human consumption, for livestock, and for fish farming. It may be the most abundant animal on the planet. Vast swarms of krill may cover almost 200 square miles of ocean and contain 2 million tons of animals. A single cubic meter of a swarm can contain 1 million animals. The peak krill catch was over 500,000 tons in 1982 and now averages about 100,000 tons per year. The estimates of total krill stocks are in the tens to several hundred millions of tons, so the harvest rate at the moment does not appear to be a threat. However, long-term studies indicate that the density of krill has dropped by 80 percent over the past seventy-five years. If that is not accounted for by harvesting, then what is happening?

Air temperature in the Antarctic Peninsula has risen by 4 to 5 degrees F in the last fifty years, more than five times the average rise across the globe. With that rise, there has been a decline in the mass of sea ice. Krill feed on algae that live on the sea ice, so it appears that the chain of dominoes is global warming leads to sea ice melting leads to less algae food leads to fewer krill.

Of course, as you are well aware, there are concerns that the rise of air and water temperature and the loss of sea ice will continue unabated with global warming. Some scientists have forecasted a rise of up to 4 to 5 degrees F in temperatures over the next 100 years. In the Antarctic, this would certainly reduce the amount of sea ice available for algal growth, and stress both animal populations that depend on krill as well as those that are cold-adapted.

What, then, is the fate of the marvelous icefish? These potential climate changes are worrisome enough, but the icefish has already experienced catastrophe. As the whaling industry died, attention turned to other Antarctic fisheries, including the icefish. The mackerel icefish *Champsocephalus gunnari* (color plate B) was fished beginning in 1971. By 1978, the catch had peaked at 235,000 tons. And then the same sad history I have recounted throughout this chapter repeated itself, with the fishery's collapse and the catch dropping to 13,000 tons in 1991, and then to only 66 tons in 1992, with no subsequent recovery.

We might have saved the blue whale from extermination (after decades of dawdling and political wrangling), but we clearly did not learn the bigger lesson.

Whatever the fate of the icefish, it is clear that its recovery will require it to swim against the current of overfishing, ecosystem disruption, and climate change. With its physiology so tuned to the subfreezing water temperature, it is not clear whether its depleted populations could adapt to a substantial rise in water temperature over a period as short as a century, compounded by the added stress of decreased food sources.

The Jarring Gong

> Want of foresight, unwillingness to act when action would be simple and effective, lack of clear thinking, confusion of counsel until the emergency comes, until self-preservation strikes its jarring gong—these are the features which constitute the endless repetition of history.
>
> —Winston Churchill,
> *May 2, 1935 speech to the House of Commons*

I realize that this chapter is ending our afterdinner conversation on a most downbeat note. But I believe this is absolutely necessary. I am a professional biologist, yet I was largely unaware of the dire statistics I have just presented until I did the research for this chapter. You can at least be thankful that I did not examine trends in the rain forests or other habitats here.

Winston Churchill's words are the most apt parting message I can offer. The future of nature at present looks terribly gloomy, much as the state of geopolitics looked to Churchill in 1935. He saw the threat of fascism, Nazism, and Soviet-style communism and issued countless warnings that went unheeded for years. Then, as now, most of the West's leaders were in denial, guided by wishful thinking and blind optimism. They made symbolic gestures in worthless treaties, empty platitudes, and spineless appeasement. After France's rapid collapse in 1940, Churchill's Britain stood alone until the United States was finally pushed into the war by Japan's surprise attack on Pearl Harbor. The war on Nature has been waged with increasing intensity over the past fifty years, but few powerful allies have come to her aid.

The challenge to halt the decline of nature, let alone to reverse its decimation, lies ahead. Ransom Myers laments, "We are in massive denial and continue to bicker over the last shrinking numbers of survivors, employing satellites and sensors to catch the last fish left. We have to understand how close to extinction some of these populations really are. And we must act now, before they have reached the point of no return. I want there to be hammerhead sharks and blue fin tuna around when my five-year-old son grows up. If present fishing levels persist, these great fish will go the way of the dinosaurs."

Of course, a common response to such concerns is the familiar denial and rhetoric of self-interest. Myers was stifled by bureaucrats when he worked in his province's Department of Fisheries and Oceans, reprimanded for disagreeing with the department when it tried to blame the cod collapse on cold water or increasing numbers of seals, and was even sued for libel when he publicly criticized the department.

Concerns about the environment are often discounted as alarmist, or merely philosophical or aesthetic sentiment. But the cold hard data obliterates any question about motives. Just consider the present statistics and past experience as a practical matter of self-interest. Ransom Myers and Boris Worm write, "Today's management decisions will determine whether we will enjoy biologically diverse, economically profitable fish communities twenty or fifty years from now, or whether we will have to look back on a history of collapse and extinction that was not reversed in time."

Yet, here we are at the opening of the twenty-first century, with the benefit of two centuries of evolutionary science, still debating the existence of evolution. And with more than two centuries of experience of the consequences of overfishing, overhunting, and pollution, chasing what few fish are left.

The Huxley brothers remind us that "facts do not cease to exist because they are ignored" and that we are now "determining the future direction of evolution on this earth." Will we heed these facts and accept our responsibility, if only out of self-interest? Or will cod, tuna, marlin, blue whales, dugongs, icefish, and more become as rare as palm trees in Wyoming?

Sources and Further Reading

The discoveries and ideas I have described are the accomplishments of a very large number of biologists. Because this book is intended for a broad audience, I chose not to name every individual or laboratory associated with each specific discovery, nor to use footnotes in the text; instead, I provide two sets of sources and references in this section. First, I offer some suggestions for books on evolution for further reading. And second, I detail a chapter-by-chapter summary of the original references upon which I relied. In most cases, the titles of scientific journal articles are omitted because the citation includes enough information for interested readers to locate them. Abstracts of most biological research papers can be accessed via a free government-sponsored database called PubMed at http://www.ncbi .nlm.nih.gov/entrez/query.fcgi.

General Books on Evolution

Of the many excellent books on evolution, the following may be of particular interest to readers of this book.

Carroll, Sean. *Endless Forms Most Beautiful: The New Science of Evo Devo and the Making of the Animal Kingdom.* New York: W. W. Norton, 2005.

Conway Morris, Simon. *Life's Solution: Inevitable Humans in a Lonely Universe.* Cambridge: Cambridge University Press, 2003.

Dawkins, Richard. *The Ancestor's Tale: A Pilgrimage to the Dawn of Evolution.* New York: Houghton Mifflin, 2004.

————. *The Blind Watchmaker: Why the Evidence of Evolution Reveals a Universe Without Design*. New York: W. W. Norton, 1986.

————. *Climbing Mount Improbable*. New York: W. W. Norton, 1996.

Desmond, Adrian, and James Moore. *Darwin: The Life of a Tormented Evolutionist*. London: Michael Joseph, 1991.

Knoll, Andrew. *Life on a Young Planet: The First Three Billion Years of Evolution on Earth*. Princeton, N.J: Princeton University Press, 2003.

Palumbi, Stephen. *The Evolution Explosion: How Humans Cause Rapid Evolutionary Change*. New York: W. W. Norton, 2001.

Ridley, Matt. *The Red Queen: Sex and the Evolution of Human Nature*. London: Penguin, 1993.

Weiner, Jonathan. *The Beak of the Finch: A Story of Evolution in Our Time*. New York: Alfred A. Knopf, 1994.

Zimmer, Carl. *Evolution: The Triumph of an Idea*. New York, HarperCollins, 2001.

Preface. Beyond Any Reasonable Doubt

The exoneration of Kevin Lee Green is mentioned in the report "What Every Law Enforcement Officer Should Know About DNA Evidence" (National Institute of Justice, National Commission on the Future of DNA Evidence, 1999) and in the case histories of the Innocence Project (www.innocenceproject.org). The increased use of DNA matching to solve cold cases is described in the "National Forensic DNA Study Report" (Smith Alling Lane, P.S., and Washington State University, 2003).

The rapid growth in the volume of DNA sequences in public databases is tracked at GenBank (www.ncbi.nih.gov/genbank/genbankstats.html).

1. Introduction: The Bloodless Fish of Bouvet Island

The voyage and work undertaken by Ditlef Rustad on the S.S. *Norvegia*, and his photographs, are documented in O. Holtedahl, ed., *Scientific Results of the Norwegian Antarctic Expeditions, 1927–1928* (Oslo: I Kommisjon Hos Jacob Dybwad, 1935).

Johan Ruud described his journeys to the Antarctic in *Scientific American* 213 (1965): 108–15. His original paper on the absence of red blood cells in the icefish is in *Nature* 173 (1954): 848–50.

Descriptions of the globin genes of icefish are reported in G. di Prisco et al., *Gene* 295 (2002): 185–91; E. Cocca et al., *Proceedings of the National Academy of Sciences, USA* 92 (1995): 1817–23; and Y. Zhao et al., *Journal of Biological Chemistry* 273 (1998): 14745–52. A general review of icefish biology and genetics is B. D. Sidell, *Gravitational and Space Biology Bulletin* 13 (2000): 25–34. The functional evolution of icefish microtubules is reported in H. W. Detrich et al., *Journal of Biological Chemistry* 275 (2000): 37038–47. The origin of icefish antifreeze is reported in L. Chen, A. L. DeVries, and C.-H. C. Cheng, *Proceedings of the National Academy of Sciences, USA* 94 (1997): 3811–16, and the fate of icefish myoglobin is detailed in B. D. Sidell et al., *Proceedings of the National Academy of Sciences, USA* 94 (1997): 3420–24; T. J. Moylan and B. D. Sidell, *The Journal of Experimental Biology* 203 (2000): 1277–86; and D. J. Small et al., *The Journal of Experimental Biology* 206 (2003): 131–39.

The geology of the Antarctic and history of the Southern Ocean is reviewed in J. Zachos et al., *Science* 292 (2001): 686–93; and ocean cooling is described in A. E. Shevenall, J. P. Kennett, and D. W. Lea, *Science* 305 (2004): 1766–70. The history of the notothenioid fishes is obtained from T. J. Near, *Antarctic Science* 16 (2004): 37–44; and T. J. Near, J. J. Pesavento, and C.-H. C. Cheng, *Molecular Phylogenetics and Evolution* 32 (2004): 881–91.

The histories of Darwin's voyage, discoveries, life, and writings are available in a number of superb biographies. For specific historical dates and details, I relied upon A. Desmond and J. Moore, *Darwin: The Life of a Tormented Evolutionist* (London: Michael Joseph, 1991). The passages from *On the Origin of Species* are taken from the first edition published in 1859. The phrase "survival of the fittest" first appeared in philosopher Herbert Spencer's *Social Statics* (1851).

The quotation from Sir Peter Medawar are taken from J. A. Moore, *Science as a Way of Knowing: The Foundation of Modern Biology* (Cambridge, Mass.: Harvard University Press, 1993).

The decline in Antarctic krill is described in A. Atkinson et al., *Nature* 432 (2004): 100–103.

2. The Everyday Math of Evolution: Chance, Selection, and Time
California State University professor Mike Orkin discussed the odds of winning the lottery relative to having a car accident or other mishaps in an

interview with CNN reporter Daryn Kagan on August 22, 2001. The odds of dog bite, shark attack, and mountain lion mauling were reported by Scott Latee in the *San Diego Union-Tribune* of February 22, 2004.

Darwin's interest in pigeons is also evident in his *The Variation of Animal and Plants Under Domestication* (London: John Murray, 1868). Information about William Castle, R. C. Punnett, H. T. J. Norton, J. B. S. Haldane, and others was drawn in part from a comprehensive and succinct history of evolutionary genetics by William B. Provine, *The Origins of Theoretical Population Genetics* (Chicago: University of Chicago Press, 1971). General surveys of natural selection in the wild include: John A. Endler, *Natural Selection in the Wild* (Princeton, N.J.: Princeton University Press, 1986); J. G. Kingsolver et al., *The American Naturalist* 157 (2001): 245–61; and A. P. Hendry and M. T. Kinnison, *Evolution* 53 (1999): 1637–53.

The evolution of the peppered moth is described in M. Majerus, *Melanism: Evolution in Action* (Oxford: Oxford University Press, 1998); B. S. Grant, *Evolution* 53 (1999): 980–84; and B. S. Grant, D. F. Owen, and C. A. Clarke, *Journal of Heredity* 87 (1996): 351–57; see also J. Mallet, *Genetics Society News*, issue 50: 34–38, and J. Coyne, *Nature* 396 (1998): 35–36.

A summary of selection coefficients operating on animal coloration is given by H. Hoekstra, K. E. Drumm, and M. W. Nachman, *Evolution* 58 (2004): 1329–41. The long-term field study of peregrine falcon predation on pigeons is by A. Palleroni et al., *Nature* 434 (2005): 973–74. The long-term study of stickleback evolution in Loberg Lake is M. A. Bell, W. E. Aguirre, and N. J. Buck, *Evolution* 58 (2004): 814–24.

Mutation rates in mammals are summarized in S. Kumer and S. Subramanian, *Proceedings of the National Academy of Sciences, USA* 99 (2002): 803–8, and the estimated human mutation rate is from M. W. Nachman and S. L. Crowell, *Genetics* 154 (2000): 297–304. The rates of mutations causing black fur color in pocket mice are estimated from estimates of natural mutation rates in the house mouse by G. Schlager and M. M. Dickie, *Mutation Research* 11 (1971): 89–96; and the number of mutable sites is from M. W. Nachman, H. E. Hoekstra, and S. L. D'Agostino, *Proceedings of the National Academy of Sciences, USA* 100 (2003): 5268–73, and references therein. The mathematical formulas for estimating times required for evolutionary change are obtained from Wen-Hsiung Li, *Molecular Evolution* (Sunderland, Mass.: Sinauer Associates, 1997).

Reproductive rates in pocket mice are taken from H. E. Hoekstra and M. W. Nachman, *University of California Publications in Zoology* 2005:61–81. The role of migration and selection in rock pocket mice populations is discussed in M. W. Nachman, *Genetica* 123 (2005): 125–36.

A general survey of rates of evolution is by P. D. Gingerich, *Science* 222 (1983): 159–61.

Kimura's classic works on the neutral theory include M. Kimura, *Nature* 217 (1968): 624–26; M. Kimura, *Proceedings of the National Academy of Sciences, USA* 63 (1969): 1181–88; and M. Kimura, *The Neutral Theory of Molecular Evolution* (Cambridge: Cambridge University Press, 1983).

3. Immortal Genes: Running in Place for Eons

T. D. Brock's discoveries and experiences in Yellowstone National Park are described in T. D. Brock, *Annual Review of Microbiology* 49 (1995): 1–28; T. D. Brock, *Genetics* 145 (1997): 1207–10; and T. D. Brock, *Life at High Temperatures* (Yellowstone Association for Natural Science, History, and Education, 1994). Carl Woese's key works on establishing the domain Archaea include C. R. Woese and G. F. Fox, *Proceedings of the National Academy of Sciences, USA* 74 (1977): 5088–90; and C. R. Woese, O. Kandler, and M. L. Wheel, *Proceedings of the National Academy of Sciences, USA* 87 (1990): 4576–79.

The comparison of sequenced genomes is now a very large literature. The data cited on specific species were obtained from E. V. Koonin, *Nature Reviews Microbiology* 1 (2003): 127–36; K. S. Makarova et al., *Genome Research* 9 (1999): 608–28; R. L. Tatusov et al., *BMC Bioinformatics* 4 (2003): 41; G. M. Rubin et al., *Science* 287 (2000); 2204–15; K. S. Marakova and E. V. Koonin, *Genome Biology* 4 (2003): 115; and O. Jaillon et al., *Nature* 431 (2004): 946–57.

The "immortal" core of proteins shared among all domains of life is discussed in E. V. Koonin, *Nature Reviews Microbiology* 1 (2003): 127–36. The estimate of roughly 500 core proteins was a personal communication of E. V. Koonin to S. B. Carroll (November 2, 2004).

The evolution of eukaryotes from archaean and bacterial "parents" is detailed in M. C. Rivera and J. A. Lake, *Nature* 431 (2004): 152–55; and discussed further by W. Martin and T. M. Embley, *Nature* 431 (2004): 134–36; and A. B. Simonson et al., *Proceedings of the National Academy of Sciences, USA* 102 (2005): 6608–13.

The sequences of elongation factor-1$\propto$ were obtained from the public database GenBank. The signature sequence of elongation factor-1α in eukaryotes and archaea is reported by M. C. Rivera and J. A. Lake, *Science* 257 (1992): 74–76.

4. Making the New from the Old

The role of color vision in primate biology is reviewed by B.C. Regan et al., *Proceedings of the Royal Society of London B Biological Sciences* 356 (2001): 229–83. The food preferences of colobus monkeys, chimpanzees, and howler monkeys are described by N. J. Dominy and P. W. Lucas, *Nature* 410 (2001): 363–66 and P. W. Lucas et al. *Evolution* 54 (2003): 2636–43. Specific fruit preferences of chimpanzees were conveyed by personal communication from N. J. Dominy to S. B. Carroll (June 14, 2005).

The importance of seeing color vision from the viewpoint of animals was articulated by R. Dalton, *Nature* 428 (2004): 596–97. Basic information about the composition of light and the mechanism of color vision is found in introductory physics and biology textbooks as well as tutorials available online—see, for instance, N. A. Campbell and J. B. Reece, *Biology*, 7th ed. (San Francisco: Benjamin Cummings, 2004). A review of color vision evolution and physiology in humans is J. Nathans, *Neuron* 24 (1999): 299–312. The prevalence of color blindness in macaque monkeys is reported by A. Onishi et al., *Nature* 402 (1999): 139–40.

The resolution of the hominid evolutionary tree using SINE elements is by A.-H. Salem et al., *Proceedings of the National Academy of Sciences. USA* 100 (2003): 12787–91.

The classic reference on the role of gene duplication in evolution is S. Ohno, *Evolution by Gene Duplication* (Berlin and New York: Springer-Verlag, 1970). Recent overviews include M. Lynch and V. Katju, *Trends in Genetics* 20 (2004): 544–49; J. A. Cotton and R. D. M. Page, *Proceedings of the Royal Society B* 272 (2005): 277–83; and M. Lynch and J. S. Conery, *Science* 290 (2000): 1151–55.

The literature on color vision in vertebrates is vast. A general overview of the molecular evolution of opsins is S. Yokoyama, *Gene* 300 (2002): 69–78. Other references used for this chapter include S. Yokoyama and F. B. Radlwimmer, *Genetics* 158 (2001): 1697–1710; J. I. Fasick and P. R. Robinson, *Visual Neuroscience* 17 (2000): 781–88; J. I. Fasick and P. R.

Robinson, *Biochemistry* 37 (1998): 433–38; S. Yokoyama and N. Takenaka, *Molecular Biology and Evolution* 21 (2004): 2071–78; and A. J. Hope et al., *Proceedings of Biological Science* 22 (1997): 155–63. The relationships of cows and other ungulates with cetaceans are resolved in M. Nikaido, A. P. Rooney, and N. Okada, *Proceedings of the National Academy of Sciences, USA* 96 (1999): 10261–66; and M. Nikaido et al., *Proceedings of the National Academy of Sciences, USA* 98 (2001): 7384–89.

The tuning of opsins in the ultraviolet is described in Y. Shi and S. Yokoyama, *Proceedings of the National Academy of Sciences, USA* 100 (2003): 8308–13; S. Yokoyama, F. B. Radlwimmer, and N. S. Blow, *Proceedings of the National Academy of Sciences, USA* 97 (2000): 7366–71; Y. Shi, F. B. Radlwimmer, and S. Yokoyama, *Proceedings of the National Academy of Sciences, USA* 98 (2001): 11731–36; and A. Ödeen and O. Håstad, *Molecular Biology and Evolution* 20 (2003): 855–61.

The functional role of ultraviolet light has been demonstrated in: zebra finches, A. T. D. Bennett et al., *Nature* 380 (1996): 433–35; starlings, A. T. D. Bennett et al., *Proceedings of the National Academy of Sciences, USA* 94 (1997): 8618–21; blue tits, S. Hunt et al., *Proceedings of the Royal Society of London B Biological Science* 265 (1998): 451–55; budgerigar, S. M. Pearn et al., *Proceedings of the Royal Society of London B Biological Science* 268 (2000); 2273–79; nestlings, V. Jaurdie et al., *Nature* 431 (2004): 262–63; and birds in general, R. Dalton, *Nature* 428 (2004): 596–97. The prevalence of UV-reflecting plumage is described in F. Hausmann et al., *Proceedings of the Royal Society of London B Biological Sciences* 270 (2003): 61–67; and in sibling species of tanagers in R. Bleiweiss, *Proceedings of the National Academy of Sciences, USA* 101 (2004): 16561–64. The role of UV vision in kestrels is described in J. Vlitala et al., *Nature* 373 (1995): 425–27; and in bats by Y. Winter, J. Lopez, and O. van Helverson, *Nature* 425 (2003): 612–14.

The evolution of pancreatic ribonuclease genes in colobus monkeys is described in J. Zhang, *Nature Genetics*, in press.

5. Fossil Genes: Broken Pieces of Yesterday's Life

The discovery of the coelacanth and the ongoing study of coelacanth biology are described in K. S. Thomson, *Living Fossil: The Story of the Coelacanth* (New York: W. W. Norton, 1991); and S. Weinberg, *A Fish Caught in Time: The Search for the Coelacanth* (New York: HarperCollins, 2000).

The fossilization of the coelacanth SWS opsin is described in S. Yokoyama et al., *Proceedings of the National Academy of Sciences, USA* 96 (1999): 6279–84. The fossilization of the bottlenose dolphin SWS opsin is described in J. I. Fasick et al., *Visual Neuroscience* 15 (1998): 643–51; and of various dolphins and whales in D. H. Levenson and A. Dizon, *Proceedings of the Royal Society of London B Biological Science* 270 (2003): 673–79.

The fossilization of the owl monkey and bush baby SWS opsin is described in G. H. Jacobs, M. Neitz, and J. Neitz, *Proceedings of the Royal Society of London B Biological Science* 263 (1996): 705–10; of the slow loris SWS opsin in S. Kawamura and N. Kubotera, *Journal of Molecular Evolution* 58 (2004): 314–21; and of the blind mole rat SWS opsin in Z. K. David-Gray et al., *European Journal of Neuroscience* 16 (2002): 1186–94.

The large repertoire of olfactory receptor genes in the mouse is described in X. Zhang et al., *Genomics* 83 (2004): 802–11. The fossilization of human olfactory receptor genes is detailed in Y. Nimura and M. Nei, *Proceedings of the National Academy of Sciences, USA* 100 (2003): 12235–40; B. Malnic, P. A. Godfrey, and L. R. Buck, *Proceedings of the National Academy of Sciences, USA* 101 (2004): 2584–589; and Y. Gilad et al., *Public Library of Science Biology* 2 (2004): 120–25. The fossilization of the pheromone pathway component is described in E. R. Liman and H. Inan, *Proceedings of the National Academy of Sciences, USA* 100 (2003): 3328–32.

The fossilization of galactose pathway genes in yeast is shown by C. T. Hittinger, A. Rokas, and S. B. Carroll, *Proceedings of the National Academy of Sciences, USA*, 101 (2004): 14144–49. The massive decay of genes in the leprosy bacillus is described by S. T. Cole et al., *Nature* 409 (2001): 1007–11.

The disruption of color flower genes in morning glories was demonstrated by R. A. Zufall and M. D. Rausher, *Nature* 428 (2004): 847–50. The disruption of the human *MYH16* was reported by H. Stedman et al., *Nature* 428 (2004): 415–18; further analysis by G. H. Perry, B. C. Verrelli, and A. C. Stone, *Molecular Biology and Evolution* 22 (2004): 379–82.

6. Déjà Vu: How and Why Evolution Repeats Itself
The discovery of full trichromatic vision is the howler monkey is reported in G. H. Jacobs et al., *Nature* 382 (1996): 156–58. The duplication of howler opsin genes is described in D. M. Hunt et al., *Vision Research* 38

(1998): 3299–3306; and K. S. Dulai et al., *Genome Research* 9 (1999): 629–38. An overview of the evolutionary history of primate color vision is D. M. Hunt, *Biologist* 48 (2001): 67–71. The correlation between the loss of olfactory receptor genes and the gain of color vision is described by Y. Gilad et al., *Public Library of Science Biology* 2 (2004): 120–25.

The evolution of ruminant ribonuclease genes is described in J. Zhang, *Molecular Biology and Evolution* 20 (2003): 1310–17; and of the repeated evolution of ribonuclease genes in colobine monkeys in J. Zhang, Y. Zhang, and H. F. Rosenberg, *Nature Genetics* 30 (2002): 411–15; and J. Zhang (TBA). The repeated loss of galactose pathway genes in fungi is reported in C. T. Hittinger, A. Rokas, and S. B. Carroll, *Proceedings of the National Academy of Sciences, USA* 101 (2004): 14144–49.

The repeated inactivation of the same pigmentation gene in Mexican blind cave fish was discovered by M. E. Protas et al., *Nature Genetics* (2005): 38:107–111. The repeated evolution of melanic forms of vertebrates through mutations in the *MC1R* gene is reviewed by M. E. N. Majerus and N. I. Mundy, *Trends in Genetics* 19 (2003): 585–88; and original reports include N. I. Mundy et al., *Science* 303 (2004): 1870–73; S. M. Doucet et al., *Proceedings of the Royal Society of London B* Biological Science 271 (2004): 1663–70; E. B. Rosenblum, H. E. Hoekstra, and M. W. Nachman, *Evolution* 58 (2004): 1794–1808; N. I. Mundy and J. Kelly, *American Journal of Physical Anthropology* 121 (2003): 67–80; E. Eizirik et al, *Current Biology* 13 (2003): 448–53; K. Ritland et al., *Current Biology* 13 (2001): 1468–72; M. W. Nachman et al., *Proceedings of the National Academy of Sciences, USA* 100 (2003): 5268–73; and E. Theron et al., *Current Biology* 11 (2001): 550–57.

The independent evolution of antifreeze in Arctic and Antarctic fish is described by L. Chen, A. L. DeVries, and C.-H. C. Cheng, *Proceedings of the National Academy of Sciences, USA* 94 (1997): 3817–22.

The different structures of potassium-channel-blocking toxins are reported in S. Gasparini, B. Gilquin, and A. Menez, *Toxicon* 43 (2004): 901–8; M. Dauplous et al., *The Journal of Biological Chemistry* 272 (1997): 4302–9; S. Gasparini et al., *The Journal of Biological Chemistry* 273 (1998): 25393–25403; and K.-J. Shon, *The Journal of Biological Chemistry* 273 (1998): 33–38.

The sequences of bird SWS opsins were obtained from GenBank. The estimates of bird population sizes are from the Audubon Society "State of the Birds 2004," *Audubon*, September–October 2004.

The book by Jacques Monod to which I refer is *Chance and Necessity* (New York: Vintage, 1971). For a detailed examination of the pervasive phenomenon of evolution repeating itself at the molecular and anatomical levels, see S. Conway Morris, *Life's Solution: Inevitable Humans in a Lonely Universe* (Cambridge: Cambridge University Press, 2003).

7. Our Flesh and Blood: Arms Races, the Human Race, and Natural Selection

The case history of the human death caused by the swallowing of an Oregon rough-skinned newt is detailed by S. G. Bradley and L. J. Klika, *Journal of the American Medical Association* 246 (1981): 247. The arms race between the newts and garter snakes is described in S. Geffeney et al., *Science* 297 (2002): 1336–39; and B. L. Williams et al., *Herpetologica* 59 (2003): 155–63.

A synopsis of W. C. Wells's life is given by W. H. G. Armytage, *British Medical Journal* 6 (1957): 1302. Two general reviews of theories concerning human skin color, geography, and natural selection are J. Diamond, *Nature* 435 (2005): 283–84; and N. G. Jablonski, *Annual Review of Anthropology* 33 (2004): 585–623. Genetic variation at the human *MC1R* gene and its role in the diversity of skin coloration and natural selection are reported by B. K. Rana et al., *Genetics* 151 (1999): 1547-1557; R. M. Harding et al., *American Journal of Human Genetics* 66 (2000): 1351–61; L. Naysmith, *Journal of Investigative Dermatology* 122 (2004): 423–28; E. Healy et al., *Human Molecular Genetics* 10 (2001): 2397–2402; and A. R. Rogers et al., *Current Anthropology* 45 (2004): 105–7.

Two firsthand accounts of Anthony Allison's quest to understand the relationship between sickle cell anemia and malaria are A. C. Allison, *Genetics* 166 (2004): 1391–99; and A. C. Allison, *Biochemistry and Molecular Biology Education* 30 (2002): 279–87. Original papers include A. C. Allison et al., *Anthropological Institute* 82 (1952): 55–60; A. C. Allison, *British Medical Journal* 1 (1954): 290–94; and A. C. Allison, *Transactions of the Royal Society for Tropical Medicine and Hygiene* 48 (1954): 312–18. J. B. S. Haldane mentioned a potential relationship between thalessemia and malaria in *Proceedings International Congress on Genetics and Heredity* 35 (1949): 267–73 (supplement). The multiple origins of the sickle cell mutation in Africa and Asia are reported by D. Labie et al., *Human Biology* 61 (1989): 479–91; J. Pagnier et al., *Proceedings of the National Academy of Sciences,*

USA 81 (1984): 1771–73; A. E. Kulozik et al., *American Journal of Human Genetics* 39 (1986): 239–44; and C. Lapouméroulie et al., *Human Genetics* 89 (1992): 333–37.

A general overview of malaria and human history is R. Carter and K. N. Mendis, *Clinical Microbiology Reviews* 15 (2002): 564–94. The role of G6PD mutations in malarial resistance is discussed by L. Luzzotto and R. Notaro, *Science* 294 (2001): 442–43; and reported by S. A. Tishkoff et al., *Science* 293 (2001): 455–62. The resistance of west Africans to *Plasmodium vivax* is described by L. H. Miller et al., *The New England Journal of Medicine* 295 (1976): 302–4. A short synopsis of malarial disease and treatment is available at the Centers for Disease Control Web site, www.cdc.gov/malaria/history/. The recent origin of the *P. falciparum* parasite is reported by S. K. Volkman et al., *Science* 293 (2001): 482–84.

The possible role of the cystic fibrosis mutation in resistance to pathogens is put forth in G. B. Pier, M. Grant, and T. S. Zaidi, *Proceedings of the National Academy of Sciences, USA* 94 (1997): 12088–93; and G. B. Pier et al., *Nature* 393 (1998): 79–82. A synopsis of the role of the CCR5 receptor in resistance to HIV is by E. de Silva and M. P. H. Stumpf, *FEMS Microbiology Letters* 241 (2004): 1–12.

The search for more cost-effective malarial drugs and control strategies is described by S. Sternberg in *USA Today*, April 28, 2004.

The role of mutation and selection in cancer progression and its relationship to the evolutionary process is described by F. Michor, Y. Iwasa, and M. A. Nowak, *Nature Reviews Cancer* 4 (2004): 197–206. The resistance of chronic myelogenous leukemias to Gleevec is reported by M. E. Gorre et al., *Science* 293 (2001): 876–80; and C. Rouche-Lestienne and C. Prudhomme, *Seminars in Hematology* 40 (2003): 80–82 (supplement). The strategy for overcoming Gleevec resistance is reported by N. P. Shah et al., *Science* 305 (2004): 399–401.

8. The Making and Evolution of Complexity

Darwin's theory of coral reef formation was outlined in full in *The Structure and Distribution of Coral Reefs* (1842). For an in-depth account of the controversy swirling around reef formation theory, see D. Dobbs, *Reef Madness: Charles Darwin, Alexander Agassiz, and the Meaning of Coral* (New York: Pantheon, 2005). I gathered information on the Great Barrier Reef geology and biology from materials and exhibits on Lady Elliot Island, Queensland, Australia.

The survey of eye evolution by L. V. Salvini-Plawen and E. Mayr is in *Evolutionary Biology* 10 (1977): 207–63. The discovery of the similarity of the *eyeless* gene to vertebrate counterparts is described by R. Quiring, et al., *Science* 265 (1994): 785–89. The ability of *Pax-6* genes to induce eye tissue is reported by G. Halder, P. Callaerts, and W. J. Gehring, *Science* 267 (1995): 1788–92. The distribution and use of the *Pax-6* gene in the animal kingdom is reviewed by W. J. Gehring and K. Ikeo, *Trends in Genetics* 15 (1999): 371–77. The development of the simple eyes of the ragworm is described by D. Arendt et al., *Development* 129 (2002): 1143–54.

Estimates of the time required to evolve a complex eye are by D. E. Nilsson and S. Pelger, *Proceedings of the Royal Society of London B Biological Science* 256 (1994): 53–58. An extensive discussion of eye types, optics, and evolution is presented by R. Dawkins in *Climbing Mount Improbable* (New York: W. W. Norton, 1996). See also R. Fernald, *Current Opinion in Neurobiology* 10 (2000): 444–50; and R. Fernald, *International Journal of Developmental Biology* 48 (2004): 701–5.

The puzzle of the origin of different photoreceptor cell types is explored in D. Arendt, *International Journal of Developmental Biology* 47 (2003): 563–71; and its apparent resolution reported in D. Arendt et al., *Science* 306 (2004): 869–71.

The general sharing of body-building tool-kit genes is described at length in my *Endless Forms Most Beautiful* (New York: W. W. Norton, 2005). The critical distinction between the evolution of physiology and form, and the different genetic mechanisms at work, is the focus of S. B. Carroll, *Public Library of Science Biology* 3 (2005): 1159–66.

The evolution of stickleback fish spine length and the role of the *Pitx1* gene is described by M. D. Shapiro et al., *Nature* 428 (2004): 717–23. The evolution of fruit fly spots is described by N. Gompel et al., *Nature* 433 (2005); 481–87; and B. Prud'homme et al., *Nature,* in press.

The significance of Darwin's use of the term "contrivance" is analyzed by R. Moore, *BioScience* 47 (1997): 107–14; and by S. J. Gould, *The Structure of Evolutionary Theory* (Cambridge, Mass.: Belknap Press, 2002).

9. Seeing and Believing

The history of childbed fever and Louis Pasteur's contributions are drawn from M. D. Reynolds, *How Pasteur Changed History: The Story of Louis Pasteur and the Pasteur Institute* (Bradenton, Fla.: McGuinn and McGuire,

1994); C. M. De Costa, *Medical Journal of Australia* 177 (2002): 668–71; P. Gallon "Découverte de l'antisepsie et de l'asepsie chirurgicale," www.char-fr.net/docs/textes/antisepsie.html (accessed 10/19/03): and C. L. Case, "Handwashing," National Health Museum, www.accessexcellenge.org/AE/AEC/CC/hand_background.html.

Richard Panek's excellent book is *Seeing and Believing: How the Telescope Opened Our Eyes and Minds to the Heavens* (New York: Penguin, 1998). Much more information about the trial of Galileo is available from many sources; see for example *The Trial of Galileo* by D. Linden (2002) at www.law.umkc.edu/faculty/projects/ftrials/galileo.

V. N. Soyfer's account of the Lysenko era is *Lysenko and the Tragedy of Soviet Science*, translated by L. Gruliow and R. Gruliow (New Brunswick, N.J.: Rutgers University Press, 1994). I also relied on Z. A. Medvedev. *The Rise and Fall of T. D. Lysenko* (New York: Columbia University Press, 1969); and H. F. Judson, *The Eighth Day of Creation: The Makers of the Revolution in Biology* (New York: Simon and Schuster, 1979). The note from Medvedev to Marshall Nirenberg is in *The Marshall W. Nirenberg Papers*, available online as part of the Profiles in Science collection at the National Library of Medicine (http://profiles.nlm.nih.gov). A short profile of N. Vavilov is by J. F. Crow, *Genetics* 134 (1993): 1–4. The state of Soviet agriculture influence is further discussed in R. W. Wheatcraft and S. G. Davies, *The Years of Hunger: Soviet Agriculture, 1931–1933* (New York: Palgrave Macmillan, 2004); and Lysenko's influence on Chinese agricultural failures is described in J. Beckers, *Hungry Ghosts: Mao's Secret Famine* (New York: Henry Holt, 1996).

Histories of chiropractors' opposition to vaccination include S. Homola, *Bonesetting, Chiropractic, and Cultism*, a 1963 book available online at www.chirobase.org; and J. B. Campbell, J. W. Busse, and H. S. Injeyan, *Pediatrics* 105 (2000): 43–50. A survey of Canadian chiropractic students' attitudes toward vaccination is reported by J. W. Busse et al., *Canadian Medical Association Journal* 166 (2002): 1531–34. See also J. W. Busse et al., *Journal of Manipulative and Physiological Therapeutics* 28 (2005): 367–73; and S. M. Barrett, *Chiropractors and Immunization* and references cited therein at www.chirobase.org.

Three recent books on the denial of evolution are E. C. Scott, *Evolution vs. Creationism: An Introduction* (Westport, Conn.: Greenwood Press, 2004); M. Pigliucci, *Denying Evolution: Creationism, Scientism, and the Nature of Science* (Sunderland, Mass.; Sinauer Associates, 2002); and M. Ruse,

The Evolution–Creation Struggle (Cambridge, Mass.: Harvard University Press).

A wealth of information about societal attitudes toward evolution, the positions of various religious denominations, statements made by individuals and organizations hostile to evolution, detailed critiques of those statements, and general background on evolutionary science is available at the National Center for Science Education (NCSE) Web site www.ncseweb.org. If the matter of the teaching of evolution in public schools concerns you, this is an organization that you should support.

Many quotes attributed in the text were obtained from the NCSE document "Setting the Record Straight: A Response to Creationist Misinformation About the PBS Series *Evolution*." The document is a response to various critiques including that of K. Cumming, "A Review of the PBS Video Series *Evolution*" (Santee, Calif.: Institute for Creation Research, 2004).

Examples of creationist critique of evolutionary science and their many errors include H. M. Morris, *Science and the Bible* (Chicago: Moody Press, 1986); H. M. Morris, "The Scientific Case Against Evolution," *Impact* no. 330 (2000), and P. Fernandes, "The Scientific Case Against Evolution" (Ph.D. thesis, Institute of Biblical Defense, 1997).

For Ken Ham, see "The 'Missing' Link to School Violence," *Creation Magazine*, April 29, 1999. Cal Thomas's editorial concerning the Pope were published by the Los Angeles Times Syndicate on October 30, 1996. Jerry Bergman invoked Hitler in an article in *Creation ex Nihilo Journal* 13 (1999): 101–11.

Steve Jones's book on evolution and his discussion of its hijacking by political causes is *Darwin's Ghost: The Origin of Species Updated* (New York: Random House, 1999). The Cobb County textbook sticker case was *J. M. Selman et al. v. Cobb County School District and Cobb County Board of Education* and the decision was rendered January 13, 2005, by Judge Clarence Cooper, U.S. District Judge for the Northern District of Georgia Atlanta Division.

The broadcast of Reverend Harries's "Thought for the Day" was on March 15, 2002, and the text is available at www.oxford.anglican.org. For another reply to creationist statements and the distinctions between theory and fact in science, see J. Rennie, *Scientific American* 287 (2002): 78–85.

Michael Behe's book on intelligent design is *Darwin's Black Box: The Biochemical Challenge to Evolution* (New York: Free Press, 1996). Ken Miller's critique is in *Creation/Evolution* 16 (1996): 36–40. See also K. R.

Miller, "The Flagellum Unspun: The Collapse of Irreducible Complexity," in *Debating Design: From Darwin to DNA*, ed. M. Ruse and W. Dembski (Cambridge; Cambridge University Press, 2004): 81–97; and H. A. Orr, "Devolution," *The New Yorker*, May 30, 2005.

The history of globin genes I discuss is constructed from G. Di Prisco et al., *Gene* 295 (2002): 185–91; R. C. Hardison, *Proceedings of the National Academy of Sciences, USA* 98 (2001): 1327–29; N. Gillemans et al., *Blood* 101 (2003): 2842–49; and R. Hardison, *Journal of Experimental Biology* 201 (1998): 1099–1117.

Judge Jones's decision is in *T. Kitzmiller et al. v. Dover Area School District et al.*, case 04CV2688, United States District Court for the Middle District of Pennsylvania. The text of the letter signed by Wisconsin and other states' clergy can be viewed at www.uwash.edu/colleges/cols/religion _science-collaboration.htm.

10. The Palm Trees of Wyoming

The history of the building of the transcontinental railroad and the magnificent sights of the American West are described in H. T. Williams, *The Pacific Tourist: Williams Illustrated Guide to Pacific RR, California and Pleasure Resorts Across the Continent* (New York: Henry T. Williams, 1876).

The discovery of the "petrified fish cut" and the geology and fossil fauna of the Fossil Butte region are described in P. O. McGrew and M. Casiliano, *The Geologic History of Fossil Butte National Monument and Fossil Basin*, National Park Service Occasional Paper 3. Further information about the monument is available at www.nps.gov/fobu and about private fossil quarry tours and collecting at Ulrich's Fossil Gallery, Fossil Station #308, Kemmerer, Wyoming, 83101 (www.ulrichsfossilgallery.com).

The evolutionary consequences of trophy hunting on bighorn sheep are described in D. W. Coltman et al., *Nature* 426 (2003): 655–58.

The importance of cod to the expansion of sea exploration and trade is recounted in M. Kurlansky, *Cod: A Biography of the Fish That Changed the World* (New York: Walker, 1997). The collapse of the northern Atlantic cod fishery is described in E. Brubaker, in *Political Environmentalism*, ed. T. Anderson (Stanford, Calif.: Hoover Institution Press, 2000): 161–210. The role of overfishing in the evolution of cod is reported by E. M. Olsen et al., *Nature* 428 (2004): 932–35; and J. A. Hutchings, *Nature* 428 (2004):

899–900. The effect of the collapse on typical fishermen is reported by T. Bartelme, in *The Post and Carrier* (Charleston, South Carolina) of June 23, 1996. The maladaptive changes in fish in response to size selection is reported by M. R. Walsh et al., *Ecology Letters* (2006): 142–148.

The general decline of large fish is reported by R. A. Myers and B. Worm, *Nature* 423 (2003): 280–283; and R. A. Myers and B. Worm, *Proceedings of the Royal Society of London B Biological Science* 360 (2005): 13–20. Quotes are taken from an interview with Myers and Worm that was published in *National Geographic News*, May 15, 2003.

The near extinction of the barndoor skate is reported by J. M. Casey and R. A. Myers, *Science* 281 (1998): 690–92. The decline of shark populations in analyzed by J. K. Bauer et al., *Science* 299 (2003): 389–92.

The role of overfishing in the collapse of coastal ecosystems is detailed in J. B. C. Jackson et al., *Science* 293 (2001): 629–38. The global decline of coral reefs is analyzed in J. H. Pandolfi et al., *Science* 301 (2003): 955–58.

The sorrowful state of Chesapeake Bay is detailed in the *2004 State of the Bay Report* (Annapolis, Md.: Chesapeake Bay Foundation, 2004). For a general overview of the impact of climate change, see C. Parmesan and H. Galbraith, "Observed Impacts of Global Climate Change in the U.S." (Pew Center on Global Climate Change, November 2004). The effect of climate change on the distribution of some marine fisheries is reported by A. L. Perry et al., *Science* 308 (2005): 1912–15. The role of selection in the management of sustainable fisheries is illustrated by D. O. Conover and S. B. Munch, *Science* 297 (2002): 94–96.

Estimates of the number of whales before commercial whaling are given by J. Roman and S. Palumbi, *Science* 301 (2003): 508–10. Data on the history of whaling was obtained from publications of the International Whaling Commission, and from Antarctic History, an online resource at www.antarcticonline.com. The decline of krill populations around the Antarctic Peninsula is described by A. Atkinson et al., *Nature* 432 (2004): 100–103. The change in Antarctic air and water temperatures and sea ice is discussed by Lloyd Peck of the British Antarctic Survey in *The Guardian/UK* (September 10, 2002). The rapid decline of the icefish fishery is documented in "Review of the State of the World Fishery Resources: Marine Fisheries," FAO Fisheries Circular No. 920 (Rome: Marine Resource Service, Fishery Resource Division, Fisheries Department, Food and Agriculture Organization, 1997).

Acknowledgments

When I was twelve or thirteen, an uncle of mine—let's call him Uncle Dick—asked me what I wanted to be when I grew up. "A biologist!" I blurted out. Uncle Dick furrowed his exceptionally long forehead and grimaced, "But there is no money in that."

Fortunately, my parents were not bothered by such practical concerns and encouraged each of their four children to pursue whatever interested us—and so we did. I have since learned that many happy people received the same parental advice that I heard. So, thanks Mom and Dad for allowing me to keep snakes, newts, salamanders, and lizards in the house, and their various squirmy and disgusting food items in the refrigerator.

The burden of putting up with my peculiar interests now falls upon my own family; without their support, encouragement, love, and sense of humor this project would be impossible and pointless. My wife, Jamie, has done far more than tolerate a distracted husband—she designed and selected key artwork for the book and worked hard to make the prose readable. My sons, Will and Patrick, have accompanied (or led) me to several of the magical places mentioned here, such as Yellowstone National Park and Fossil Butte National Monument. The shouting match of my stepson, Chris, with howler monkeys on our visit to Costa Rica inspired the story in chapter 6.

I also thank my siblings, who have always supported me in every endeavor. My brothers, Peter and Jim, helped shape a couple of chapters and my sister, Nancy, has continued our decade-long running discussion of the history and insights of evolutionary biology's pioneers.

I am very grateful for the generous creative and critical contributions of colleagues at the University of Wisconsin–Madison. Leanne Olds composed or adapted most of the schematic figures; Professor James F. Crow, Drs. Antonis Rokas, Benjamin Prud'homme, and Steve Paddock, and Chris Hittinger read the entire manuscript and provided detailed critiques and suggestions.

The discoveries described in the book are the fruits of the creativity and hard work of thousands of scientists who have contributed to the invention of the methods used to decipher the DNA record and who have analyzed many species' genes and genomes. I thank the many scientists who provided illustrations of their work and shared their expertise and ideas. I especially want to thank Michael Nachman, Michael Lynch, Cliff Tabin, and Eugene Koonin for detailed discussions and information.

I am also particularly indebted to the individuals who have worked with and collaborated with me over the past two decades. I have continued to learn more from my postdocs and students than I teach in return. The dedication and efforts of my longtime staff have made our research laboratory an exciting and fun place to be every day. And my close colleagues around the world have been an endless source of inspiraton and enlightenment. I have enjoyed an unusual degree of freedom in pursuing research ideas thanks to the generous support of the Howard Hughes Medical Institute.

I am especially grateful to my agent, Russ Galen, who provided crucial guidance at the conception of this book as well as great encouragement throughout its development, and to my editor, Jack Repcheck, for his infinite enthusiasm, critical input, and confidence in the importance of this book.

Index

Page numbers in *italics* refer to illustrations.